ECONOMICS AND BIOLOGICAL DIVERSITY:

Developing and Using Economic Incentives to Conserve Biological Resources

By
Jeffrey A. McNeely

INTERNATIONAL UNION FOR CONSERVATION OF
NATURE AND NATURAL RESOURCES
Gland, Switzerland
November 1988

Prepared and published by the International Union for Conservation of Nature and Natural Resources

Citation: McNeely, Jeffrey A. 1988. *Economics and Biological Diversity: Developing and Using Economic Incentives to Conserve Biological Resources*. IUCN, Gland, Switzerland. xiv + 232 pp.

ISBN: 2-88032-964-7

Design: The Magazine Group

Printed by: St. Mary's Press, McGregor and Werner, Inc. Washington, D.C.

Available from: IUCN Publications Services, 1196 Gland, Switzerland

The designations of geographical entities in this book, and the presentation of the material, do not imply the expression of any opinion whatsoever on the part of the participating organizations concerning the legal status of any country, territory, or area, or of its authorities, or concerning the delimitation of its frontiers or boundaries.

The interpretations and conclusions in this report are those of the author and do not necessarily represent the view of IUCN, USAID, or the other participating organizations.

TABLE OF CONTENTS

PREFACE

The World Commission on Environment and Development has
pointed out that "Conservation of living natural resources—
plants, animals, and micro-organisms, and the non-living elements
of the environment on which they depend—is crucial for development.... The challenge facing nations today is no longer deciding
whether conservation is a good idea, but rather how it can be implemented in the national interest and within the means available in
each country" (WCED, 1987).

The WCED called for greater attention by development agencies to the issues of conserving biological diversity, suggested that
national accounting systems should incorporate the value of biological resource stocks, and advocated a series of economic measures
to support conservation. The International Union for Conservation
of Nature and Natural Resources (IUCN) has taken up these challenges by attempting to design specific mechanisms that governments, non-governmental organizations (NGOs), and development
assistance agencies can use to promote the conservation of biological diversity. One mechanism which has received insufficient attention by conservation organizations has been economic incentives
and disincentives.

This document grew out of workshops held in San Jose, Costa
Rica, on 4–5 February 1988, and Washington D.C. on 13 April 1988.
At these workshops, case studies were presented, concepts were
discussed, and general approaches were agreed. The document,

This paper owes a particular debt to Scott Barrett, who prepared the framework paper for the San José workshop and has provided considerable technical support to the entire economics work of IUCN. The paper has also benefited greatly from discussions held with Robert Goodland, John Kru·illa George Ledec, Ernst Lutz, John MacKinnon, Jim MacNeill, Dorothy Marschak, David Pearce, Robert Repetto, C. Ford Runge, Jerry Warford, and Mike Wells. Useful comments were also received from Bob Dobias, Patrick Durst, Adrian Phillips, Robert Prescott-Allen, and Christine Prescott-Allen.

Staff at IUCN Headquarters who commented on earlier drafts include: Frédéric Briand, Francoise Burhenne, Michael Chilcott (WWF), Pat Dugan, Chris Elliot (WWF), Danny Elder, Vitus Fernando, Michael Green, Mark Halle, Martin Holdgate, Dan Navid, Per Ryden, Jeff Sayer, Simon Stuart, and Paul Wachtel (WWF). Joanna Erfani provided secretarial support, and Morag White made a number of useful editorial suggestions. My thanks goes to all of them, but the responsibility for any errors remaining rests with me.

EXECUTIVE SUMMARY

INTRODUCTION

Some of our planet's greatest wealth is contained in natural forests, plains, mountains, wetlands and marine habitats. These biological resources are the physical manifestation of the globe's biological diversity, which simply stated is the variety and variability among living organisms and the ecological complexes in which they occur. Effective systems of management can ensure that biological resources not only survive, but in fact increase while they are being used, thus providing the foundation for sustainable development and for stable national economies.

But instead of conserving the rich resources of forest, wetland, and sea, current processes of development are depleting many biological resources at such a rate that they are rendered essentially nonrenewable. Experience has shown that too little biological diversity will be conserved by market forces alone, and that effective government intervention is required to meet the needs of society. Economic inducements are likely to prove the most effective measures for converting over-exploitation to sustainable use of biological resources.

ECONOMIC OBSTACLES TO CONSERVATION

The fundamental constraint is that some people earn immediate benefits from exploiting biological resources without paying the full social and economic costs of resource depletion; instead, these costs (to be paid either now or in the future) are transferred to society as a whole. Further, the nations with the greatest biological diversity are frequently those with the fewest economic means to implement conservation programs. They need to use their biological resources to generate income for their growing populations, but problems arise when these resources are abused through misman-agement rather than nurtured through effective management.

Other major economic obstacles to conservation include:

- biological resources are often not given appropriate prices in the marketplace;
- because the social benefits of conserving biological resources are often intangible, widely spread, and not fully reflected in market prices, the benefits of protecting natural areas are in practice seldom fully represented in cost-benefit analysis;
- the species, ecosystems, and ecosystem services which are most over-exploited tend to be the ones with the weakest ownership;
- the discount rates applied by current economic planning tend to encourage depletion of biological resources rather than conservation; and
- conventional measures of national income do not recognize the drawing down of the stock of natural capital, and instead consider the depletion of resources, i.e., the loss of wealth, as net income.

ASSESSING THE VALUE OF BIOLOGICAL DIVERSITY

In order to compete for the attention of government decision-makers, conservation policies first need to demonstrate in economic terms the value of biological diversity to the country's social and economic development. Approaches for determining the value of biological resources include:

- assessing the value of nature's products—such as firewood, fodder, and game meat—that are consumed directly, with-out passing through a market;
- assessing the value of products which are commercially har-vested, such as timber, fish, ivory, and medicinal plants;
- and assessing indirect values of ecosystem functions, such as watershed protection, photosynthesis, regulation of climate, and production of soil.

Some biological resources can be easily transformed into revenue through harvesting, while others provide flows of services which do not carry an obvious price-tag. However, an ecosystem which has been depleted of its economically-important species or a habitat which has been altered to another use cannot be re-built out of income. The costs of re-establishing forests or reversing the processes of desertification can far exceed any economic benefits from over-harvesting or otherwise abusing biological resources, so the environmental costs of depletion need to be estimated in terms of the time and effort required to restore resources to their former productivity.

Assessing values and costs of protecting biological resources provides a basis for determining the total value of any protected area or other system of biological resources. Since the value of conserving biological resources can be considerable, conservation should be seen as a form of economic development. And since biological resources have economic values, investments in conservation should be judged in economic terms, requiring reliable and credible means of measuring the benefits of conservation.

USING ECONOMIC INCENTIVES TO PROMOTE CONSERVATION

To the extent that resource exploitation is governed by the perceived self-interest of various individuals or groups, behavior affecting maintenance of biological diversity can best be changed by providing new approaches to conservation which alter people's perceptions of what behavior is in their self-interest. Since self-interest today is defined primarily in economic terms, conservation needs to be promoted through the means of economic incentives.

An **incentive** for conservation is any inducement which is specifically intended to incite or motivate governments, local people, and international organizations to conserve biological diversity. A **perverse incentive** is one which induces behavior which depletes biological diversity. A disincentive is any inducement or mechanism designed to discourage depleting of biological diversity. Together, **incentives and disincentives provide the carrot and the stick for motivating behavior that will conserve biological resources.**

Direct incentives—either in cash or in kind—are applied to achieve specific objectives, such as improving management of a protected area. **Indirect incentives** do not require any direct budgetary appropriation for biological resource conservation, but apply fiscal, service, social, and natural resources policies to specific conservation problems.

Incentives are used to divert land, capital, and labor towards conserving biological resources, and to promote broader participation in work which will benefit these resources. They can smooth the uneven distribution of the costs and benefits of conserving biological resources, mitigate anticipated negative impacts on local people by regulations controlling exploitation, compensate people for any extraordinary losses suffered through such controls, and reward the local people who assume externalities through which the larger public benefits. Incentives are clearly worthwhile when they stimulate activities which conserve biological resources, at a lower economic cost than that of the economic benefits received.

To function effectively, incentives require some degree of regulation, enforcement, and monitoring. They must be used with considerable sensitivity if they are to attain their objectives, and must be able to adapt to changing conditions.

THE PROBLEM OF PERVERSE INCENTIVES

Economic incentives have been far more pervasive in overexploiting biological resources than conserving them. In most parts of the tropics, the opening of forest areas is supported by powerful economic incentives such as state-sponsored road-building programs which facilitate access to markets. Further, resettlement of poor people in the remote forested areas made accessible by new roads is often politically preferable to genuine land reform which involves the redistribution of existing agricultural lands. Governments have often instituted these perverse incentives for important political or social reasons, and the impact on the environment is often an external factor.

While incentives to convert forests and other wilderness to agricultural uses may have been appropriate when biological resources were plentiful, the process is reaching its productive limits (and indeed has exceeded them in many places). A major step in moving from exploitation to sustainable use is for governments to review the impacts of all relevant policies on the status and trends of biological resources. Based on the policy review, governments should eliminate or at least reduce policy distortions that favor environmentally unsound practices, discriminate against the rural poor, reduce economic efficiency, and waste budgetary resources. Overcoming the damage caused by perverse incentives will require new incentives to promote conservation, applied at a series of levels and in a number of sectors.

APPLYING INCENTIVES AT THE COMMUNITY LEVEL

The specific package of available biological resources varies considerably from place to place, depending on such factors as soil, rainfall, and history of human use. For the people living in or near the forests, plants and animals provide food, medicine, hides, building materials, income, and the source of inspiration; rivers provide transportation, fish, water, and soils; and coral reefs and coastal mangroves provide a permanent source of sustenance and building materials.

Depending on these resources, rural people have often developed their own means of managing a sustainable yield of benefits. Biological resources are often under threat because the responsibility for their management has been removed from the people who live closest to them, and instead has been transferred to government agencies located in distant capitals. But the costs of conservation still typically fall on the relatively few rural people who otherwise might have benefitted most directly from exploiting these resources. Worse, the rural people who live closest to the areas with greatest biological diversity are often among the most economically disadvantaged—the poorest of the poor.

Under such conditions, the villager is often forced to become a poacher, or to clear national park land to grow a crop. Changing this behavior requires first examining government resource management policies to determine how they may stimulate a villager's poaching and encroachment. Economic incentives designed to reverse the effects of these policies may provide the best means of transforming an exploiter into a conservationist.

Appropriate measures may include assigning at least some management responsibility to local institutions, strengthening community-based resource management systems, designing pricing policies and tax benefits which will promote conservation of biological resources, and introducing a variety of property rights and land tenure arrangements. These measures may serve to rekindle traditional ways and means of managing biological resources which have been weakened in recent years.

Which members of a population have their access to biological resources enhanced and which members have it restricted by government policies is of profound importance in determining whether the resources will make a sustainable contribution to society. People living in and around the forests, wetlands, and coastal zones, rather than governments, often exercise the real power over the use of the biological resources, so they should be given incentives to manage these resources sustainably at their own cost and for their own benefit.

SUPPORT FROM THE NATIONAL LEVEL FOR COMMUNITY-BASED INCENTIVES

The biological resources which support the community are also of considerable interest to the nation and the world. Further, incentives at the local community level are likely to require considerable support from compatible policies at the national level. Biological resources do not occur only in wilderness, and economic incentives may also be used more generally throughout the country to encourage settlement patterns and production systems that are directed at the sustainable use of the resources of forest, savanna, wetland, and sea. The specific policies required at the national level will derive from what is required at the community level to conserve biological resources.

Sustainable development requires coordination among a number of policies and levels. This is not as easy as it sounds. Many conservation problems are due to divided responsibility among sectoral units, leading to fragmentation, poor coordination, conflicting directives, and waste of human and financial resources. This can only be overcome by examining the impact of decisions in one sector on the ability of another sector to depend on the same resources. In most cases, the optimal balance point where the benefit of considering secondary impacts is overtaken by the cost of doing so lies well beyond the current practice of taking decisions based on a very narrow range of sectoral considerations.

INTERNATIONAL SUPPORT FOR INCENTIVES PACKAGES

Biological diversity is a public good, and species and ecosystems in one part of the world can provide significant benefits to distant nations. Indeed, some experts believe that far greater benefits from conserving native gene pools, especially in the wilds of the tropics, will be gained by wealthy temperate nations than the often poverty-stricken nations doing the conservation. Further, much of the depletion of biological diversity over the past 400 years or so has been caused by powerful global forces, primarily driven by markets in colonial, and then industrial, countries. Because the international community as a whole benefits from conservation, it should contribute to the costs of conserving biological resources.

An important means for doing so is through the provision of economic incentives from the temperate nations to the tropical ones. These can include direct incentives such as grants, loans, subsidies, debt swaps, and food; and indirect incentives such as commodities agreements, technical assistance, equipment, and information. Development assistance often contains a package of such incentives, including both direct on-the-ground projects and very abstract incentives such as peer pressure and public image.

FUNDING FOR CONSERVATION INCENTIVES

Governments seldom have sufficient capital or labor to manage their nation's biological resources in an optimum way, even though investments in conservation can be very cost-effective. Conservation programs are usually implemented through resource management agencies who need sufficient and reliable sources of support to implement an effective incentives scheme. Support from government budgets might include national bank loans, initial contribution to revolving funds, the government portion of shared costs, and education and training.

Some incentives involve little more than an administrative decision or regulation, such as the enactment of a law or monetary policy action, while others involve bilateral agreements or cooperation with international agencies, as in food for work programs. In many developing countries, large externally-supported development projects can often include elements which support incentives for conserving biological resources. Community development activities may already be in progress in communities located near areas important for conserving biological resources, in which case linkages with changed behavior toward conservation can be incorporated with little additional cost.

Additional innovative funding mechanisms for supporting incentives include: tax deductability for donations of cash, land, or services; charging entry fees to protected areas; returning profits from exploiting biological resources to the people living in the region; implementing water use charges for the water produced by a protected area; building conditionality into extractive concession agreements; seeking support from international conservation organizations; and considering "conservation concessions," similar to those for forestry or mining.

The threats to biological resources have such profound implications for humanity that governments must take decisive action, and accept that some additional investments will be required. But sustainble development of biological resources will likely be far less expensive than rehabilitation programs, and most conservation efforts have proven cost-effective on traditional economic grounds.

GUIDELINES FOR USING INCENTIVES TO CONSERVE BIOLOGICAL DIVERSITY

Action is required at the strategic level, where governments establish national and international objectives for addressing on a broad front the fundamental problems of degradation of biological resources, and at the tactical level, with specific actions designed to address specific problems. Guidelines are presented to stimulate the greatest possible government commitment to conserving the entire spectrum of biological diversity, in an economically optimal way; and to assist development agencies—both national and international—in improving the design of projects that affect biological diversity. They provide practical advice for the formulation of policies for the sustainable development of biological resources, and for the conversion of policy into practice through specific project interventions. They include detailed advice on how incentives packages can be designed and implemented by resource management agencies, and how specific project interventions can be most effective.

CHAPTER ONE

ECONOMICS AND BIOLOGICAL DIVERSITY

INTRODUCTION

Viewed from the air, the tropical forest of Brazil, Indonesia, or Zaire is a vast carpet of green, broken by occasional village clearings, rivers, and hills. These forests and waters support a great diversity of species and ecosystems, including some of our planet's greatest natural wealth. Various characteristics of these natural habitats become resources when humans begin to appreciate their potential utility, but problems have arisen as governments and local populations have increased their demands on these resources, sometimes exploiting them at rates which cannot be sustained and which are costly to society at large.

The resources of forests, savannas, and seas fall into several broad categories. Economists distinguish **non-renewable natural resources** such as oil, coal, gold, and iron from **renewable resources** such as forests, animals and grasslands; the renewable resources are inexhaustable when managed appropriately. Both non-renewable and renewable resources can be privately, communally, or governmentally owned and managed. They are also generally recognized to have market value, although market values do not always reflect their true scarcity or aesthetic value to society.

Much more difficult for economists and resource managers to deal with are **environmental resources**, which are "public goods" based on the functioning ecosystem; these include such things as the provision of clean air, functioning watersheds, biological diversity, and scenic beauty. While these environmental resources provide valuable services to people, such as the regulation of climate, support of economically important species, and formation of soil, they seldom have market prices assigned to them (Smith, 1988).

This paper will focus mainly on the renewable natural resources and environmental natural resources with important public goods characteristics; together, these can be considered **"biological resources,"** being based on genes, species, and ecosystems which have actual or potential value to people. These biological resources are the physical manifestation of the globe's **biological diversity,** which simply stated is the variety and variability among living organisms and the ecological complexes in which they occur (Box 1).

As the non-renewable resources are gradually consumed, the renewable biological resources are likely to increase in importance and nations which have maintained their rich endowments of biological diversity may well have a significant advantage over those whose biological resources have been depleted. A fundamental point to bear in mind is that **effective systems of management can ensure that biological resources not only survive, but in fact increase while they are being used, thus providing the foundation for sustainable development.**

A particular challenge comes from the fact that the areas with the greatest biological diversity are frequently those with the fewest economic means to implement conservation programs. Most of the biologically richest nations have low per capita income (compare Zaire's $160, Burma's $180, and Indonesia's $560 with the $14,070 of the USA and the $16,340 of Switzerland); and within most countries, the greatest biological diversity tends to be found in the most remote regions, where habitats are least affected by modern influences. For these biologically rich but economically poor nations and regions, using their resources to generate income for their (typically increasing) populations has first priority. Problems arise when these resources are abused through mismanagement rather than nurtured through effective management.

Since future consumption depends to a considerable extent on the stock of natural capital, conservation may well be a precondition for economic growth. Conservation is certainly a precondition for **sustainable development**, which unites the ecological concept of carrying capacity with the economic concepts of growth and development. But instead of conserving the rich resources of

Box 1. What is Biological Diversity?

Biological diversity is an umbrella term for the degree of nature's variety, including both the number and frequency of ecosystems, species, or genes in a given assemblage. It is usually considered at three different levels, **"genetic diversity," "species diversity,"** and **"ecosystem diversity."** Genetic diversity is a concept of the variability within a species, as measured by the variation in genes (chemical units of hereditary information that can be passed from one generation to another) within a particular species, variety, subspecies, or breed. Species diversity is a concept of the variety of living organisms on earth, and is measured by the total number of species in the world (variously estimated as from 5 to 30 million or more, though only about 1.4 million have actually been described), or in a given area under study.

In general, the larger the population size of a species, the greater the chance of there being high genetic diversity. But population increase in some species may lead to a population decline in other species, and even to a reduction in species diversity. Since it is usually not possible to have both maximum species diversity and maximum genetic diversity, national policy-makers should define the optimum biological diversity consistent with their development objectives; one key element is to ensure that no species falls below the minimum critical population size at which genetic diversity is lost rapidly.

Ecosystem diversity relates to the diversity and health of the ecological complexes within which species occur. Ecosystems provide natural cycles of nutrients (from production to consumption to decomposition), of water, of oxygen and carbon dioxide (thereby affecting the climate), and of other chemicals like sulphur, nitrogen, and carbon. Ecological processes govern primary and secondary production (i.e., energy flow), mineralization of organic matter in the soils and sediments, and storage and transport of minerals and biomass. Efforts to conserve species must therefore also conserve the ecosystems of which they are a part.

(Sources: OTA, 1987; Ricklefs, Naveh and Turner, 1984)

Box 2: Where Biological Diversity is Found and
How it is Conserved

Wild biological diversity is not spread evenly across the planet. In general, well-watered lowland tropical terrestrial ecosystems have the greatest diversity, with diversity declining along with rainfall and latitude (or elevation); islands or small areas of habitat tend to have fewer species than large areas of the same habitat type. On the other hand, isolated islands tend to have high degrees of endemism (species which are found nowhere else), so conserving the entire range of the world's biological wealth requires action in both centers of endemism and areas of high biological diversity.

Human influences tend to reduce diversity, particularly where they are intensive and long-standing (as in permanent agriculture), but limited human activities can actually increase diversity (as in some systems of shifting cultivation at low human population densities). Aquatic habitats parallel these generalizations, with the tropical systems—especially coral reefs and large old lakes (as in the African Rift Valley lakes)—having greater diversity than temperate systems.

Within these broad trends, some areas are more important than others, due to such factors as complexity of soils and other geological factors; altitudinal variation (areas with considerable variation in elevation containing greater diversity and being better able to adapt to climate change); and history (some areas having served as "refugia" during drier or cooler periods). Based on such factors, areas of particular importance have been assessed for tropical Africa (IUCN, 1986b), Oceania (IUCN, 1986a), and tropical Asia (MacKinnon and MacKinnon, 1986).

It is apparent that most diversity tends to be found in extensive tropical habitats which are little affected by humans, so relatively large protected areas are likely to be the most effective way of conserving maximum biological diversity (Soulé and Wilcox, 1980). But the real situation is far more complex than that, because diversity also occurs in managed forests, secondary forests, and agroecosystems. Conserving biological resources therefore requires a wide range of management tools, varying from complete protection to intensive management.

Box 2 (continued)

Technologies aimed at maintaining ecosystems include protected areas, land-use planning, zoning systems, and regulations on permissible activities (MacKinnon *et al.*, 1986); technologies aimed at managing wild species in their natural habitats include controls on harvesting or trading, enhancement of stocking rates, and habitat manipulation (Giles, 1971). All of these require research and monitoring to ensure that the technologies are effective. In addition, various off-site *(ex situ)* techniques are available, including: captive breeding or propagation programs in zoos, botanic gardens, hatcheries, and game farms; seed and pollen banks; microbial culture collections; and tissue culture collections (OTA, 1987). The latter are most suitable for maintaining diversity of agricultural species and varieties. This document will concentrate on the various on-site technologies.

forest, wetland, and sea, current processes of development are depleting many biological resources at such a rate and reducing them to such low population levels that they are rendered essentially non-renewable.

Development agencies are becoming concerned about the depletion of these species and ecosystems, with the growing awareness that development depends on their maintenance. The over-exploitation of biological resources is providing the major new development challenge of the late 20th Century. How can the process of change be managed so that biological resources can make their best contribution to sustainable development? Which economically attractive land uses are compatible with the conservation of biological diversity? What economic incentives are available to promote conservation instead of over-exploitation?

In seeking answers to such questions, those responsible for planning and implementing the process of sustainable development already have sufficient technology to manage these resources far better than is being done today. Ample guidelines exist for the management of biological resources (see, for example, Schonewald-Cox, *et al.*, 1983; MacKinnon, *et al.*, 1986; OTA, 1987), but political will has been insufficient to ensure the effective implementation of these guidelines.

The fundamental problem is that more people earn greater immediate benefits from exploiting biological resources than they do from conserving them. To the extent that resource exploitation is governed by the perceived self-interest of various individuals or groups, **behavior affecting maintenance of biological diversity can best be changed by providing new approaches to conservation which alter people's perceptions of what behavior is in their self-interest.**

Since self-interest today is defined primarily in economic terms and conserving biological diversity is part of the process of sustainable development, the decision-makers with the appropriate power and resources to influence the development process—statesmen, senior civil servants, planners, corporate directors, development assistance agencies, forest-based enterprises, and so on—are most likely to generate enthusiasm for policies which promote conservation through the means of economic inducements.

AN APPROACH TO USING ECONOMIC INCENTIVES TO PROMOTE CONSERVATION OF BIOLOGICAL DIVERSITY

The purpose of this paper is to enlist the help of economic analytic methods and policy tools to promote the conservation of optimum biological diversity in support of sustainable development. It approaches this task first by discussing how the value of biological resources can be determined, so that governments and skeptical consumers can be convinced that these resources are worth conserving. Second, it presents the types of economic incentives and disincentives which governments can use to influence resource use by participants in development activities—incentives which may lead to overexploitation or to conservation at the local, national, and international community levels. The paper concludes with guidelines for central governments, resource management agencies, and for those interested in building economic incentives into the design, implementation, and evaluation of development projects.

Since reality is far more complex than any guidelines can be, a series of case studies is presented to illustrate how economic incentives and disincentives have actually been applied to solving real-life problems in both tropical and temperate settings.

Purpose of the Guidelines

The guidelines presented in Chapter 8 are designed to promote the survival of the optimum biological diversity, and to suggest ways and means for ensuring that biological resources make their most useful contribution to sustainable development. The objectives of the guidelines include the following:

- to provide mechanisms by which biological resources can continue to support the process of sustainable development.
- to assist those who are designing, implementing, or evaluating projects which affect biological diversity to incorporate appropriate economic incentives into their projects.
- to provide all agencies concerned with biological diversity— including international organizations, development agencies, government agencies, and non-governmental organizations (NGOs)—with guidelines on how to incorporate economic methods into their efforts to conserve biological diversity.
- to help generate additional funding to supplement dwindling public funds for government and private agencies involved in conservation of biological diversity.
- to stimulate the creation of ways and means by which conservation of biological resources can be essentially self-financing (especially for key protected areas).

Limitations

Biological resources support development in virtually all sectors, and affect those who live in cities as well as in the countryside. However, this document addresses only the rural dimensions of the problem, leaving the (perhaps more difficult) problems of the urban setting to others. Further, it concentrates on wild (or "natural") biological resources and gives relatively little direct attention to agricultural issues in the belief that these issues are already being sufficiently well addressed by FAO, IBPGR, and other agencies.

Agriculture, forestry, and fisheries, managed well, are exercises in sustainable management of modified ecosystems to yield what humans perceive as optimal productivity. These systems are inevitably somewhat impoverished, as predators and competitors are eliminated or reduced and the population structure is altered in order to enhance yields; but they are ecologically sound and essential to human welfare. Such systems both affect and depend on the more natural ecosystems discussed in the following pages to ensure their long-term productivity.

It is apparent that conservation of biological diversity requires appropriate government policies in many sectors. In developing such policies, economic approaches can help clarify issues and indicate costs and benefits of alternative courses of action, but decisions about allocation of resources are perhaps even more dependent on the political and social objectives of the nation involved. Economics is therefore just one important tool among many that are available to concerned governments, and most resource problems require a variety of tools and ingredients to build the most efficient solutions.

The guidelines presented in this document are intended to provide practical advice for the formulation of policies for the sustainable development of biological resources, and for the conversion of policy into practice through specific project interventions. They cannot provide definitive answers to every situation, because each setting has its own characteristics. Factors which will affect how economic incentives and disincentives are applied in a particular case include:

- the specific nature of the local or national economy;
- the number, size, and influence of factors depleting biological resources;
- the nature of the biological resource and its response to disturbance and exploitation;
- the relative strength of local institutions;
- the technical alternatives available to counteract depletion of biological resources; and
- the authority of the control agency.

Human decision-making is inevitably based on economic thinking, irrespective of whether it is labeled as such. This document aims to demonstrate the benefits of linking economics more explicitly with the conservation of biological diversity.

CHAPTER TWO

VALUES AND BENEFITS OF BIOLOGICAL DIVERSITY

INTRODUCTION

Experience has shown that too little biological diversity will be conserved by market forces alone, and that effective government intervention is required to meet the needs of society. Unfortunately, current government policies often exacerbate the natural tendency for biological resources to be over-exploited, so new policies need to be developed to correct for the inherent failure of the market to conserve sufficient diversity.

In order to compete for the attention of government decision-makers, policies regarding biological diversity first need to demonstrate in economic terms the value of biological diversity to the country's social and economic development. Some have argued that biological resources are in one sense beyond value because they provide the biotic raw materials that underpin every major type of economic endeavor at its most fundamental level (Oldfield, 1984). But ample economic justification can be marshalled by those seeking to exploit biological resources, so the same kinds of reasoning need to be used to support alternative uses of the resources. In order for governments to assess the priority they will give to conservation of biological diversity, they need to have a firm indication of what contribution biological resources make to their national economy.

It is important to note that "conservation" does not mean non-use, but rather wise use which contributes to sustainable development.

As defined by IUCN, conservation is "the management of human use of the biosphere so that it may yield the greatest sustainable benefit to present generations while maintaining its potential to meet the needs and aspirations of future generations. Thus conservation is positive, embracing preservation, maintenance, sustainable utilization, restoration, and enhancement of the natural environment" (IUCN, 1980).

Conservation of biological diversity should therefore be seen as a form of economic development. And since biological resources have economic values, investments in conservation should be judged in economic terms, requiring reliable and credible means of measuring the benefits of conserving biological diversity (in other words, measuring the advantageous consequences or improved conditions resulting from conservation action).

THE ECONOMIC ROOTS OF OVER-EXPLOITATION

Before seeking economic tools to support conservation, it is worthwhile to review briefly why current economic systems have often led to over-exploitation of biological resources (see Clark, 1973a; Dasgupta, 1982; Fisher, 1981b; Norgaard, 1984; Pearce, 1976; and Randall, 1979 for more detailed discussions). Clearly, different types of biological resources suffer from different problems; open access fisheries, tropical forests, and land suitable for agriculture have different economic characteristics and need to be treated in different ways. However, six major issues are of concern here.

First, biological resources are often not given appropriate prices in the marketplace. Even where a biological resource is traded directly in the market, it may have associated values which are not reflected in its price. Further, the benefits of the existence of any given level of biological diversity are conferred on all who value them, and the diversity enjoyed by one individual does not reduce the amount available to others. Biological diversity is therefore a "public good," and individuals and industries can often gain its benefits without paying for them (the "free rider" problem). The often-intangible and widespread costs of depleting biological diversity usually provide ineffectual justification for conservation when balanced against projected monetary benefits of exploitation (which typically accrue to relatively few individuals).

Second, because the social benefits of conserving biological resources are often intangible, widely spread, and not fully reflected in market prices, the benefits of protecting natural areas are in practice seldom fully represented in cost-benefit analysis. In contrast, the benefits of exploiting the resources supported by natural areas are often easily measured. Hence, cost-benefit analyses usually underestimate the net benefits of conservation or, equivalently, over-estimate the net benefits of the exploitation alternative. As Old-field (1984) puts it, "Developments are proposed, the development alternatives are evaluated, the social costs of habitat losses or extinction are ignored or casually considered, and the decision to develop is given the go-ahead, actually on the basis of incomplete economic information. It is by this gradual process of land conversion that entire ecosystems and wildlife species have disappeared." In short, today's land use patterns are determined primarily by the rent-producing capacity of the area in question, irrespective of its value to society in a more natural state.

Third, those who benefit from exploiting a forest, wetland, or coral reef seldom pay the full social and economic costs of their exploitation; instead, these costs (to be paid either now or in the future) are transferred to society as a whole, or to individuals and institutions who had gained little benefit from the original exploitation. Such "external costs" are often accidental side-effects of development projects, so the loss is not recognized in either private or social cost-benefit analyses. Timber concessionaires, for example, do not need to concern themselves with the downstream siltation they are causing, or the species they are depleting, because they do not pay the full cost of these effects. Once they have logged "their" forest, they will leave, and the downstream farmer will have to pay for the siltation damage and the nation or world at large for the reduction in biological diversity. It may well be that the greatest cause of the reduction in global biological diversity is inadvertence, an external cost of the more direct financial justification for harvesting certain biological resources.

Fourth, the species, ecosystems, and ecosystem services which are most over-exploited tend to be the ones with the weakest ownership. Many of these are open access resources for which the traditional control mechanisms have failed in the face the growing demands of centralized government, national development, international trade, and population growth. Within modern and centralized systems of administration, the forests and the wildlife they contain are often publicly-owned resources which are not valued at market rates, but rather are treated as free commodities for exploitation by concessionaires. Generally speaking, the more well-defined, secure, and exclusive are the property rights to biological resources, the more effectively can the use of these resources be allocated by markets. When ownership rights are weakly enforced (either by the government or by a private owner), exploitation is allocated not to those who value the resource most, but rather to those who can pay the most for the exploitation rights. In a market situation characterized by central government control over resource use and high consumer demand, the costs of protecting species and ecosystems

---

Box 3. The Major Threats to Biological Diversity

In seeking ways and means to use economic methodology to support conservation of biological resources, it is necessary to have a clear understanding of the major threats which biological resources face on the ground and in the water. It can be seen that most of these threats have an economic foundation. Major threats include:

- Habitat alteration, usually from highly diverse natural ecosystems to far less diverse (often mono-culture) agro-ecosystems. This is clearly the most important threat, often related to land-use changes on a regional scale which involve great reduction in the area of natural vegetation; such reductions in area inevitably mean reductions in populations of species, with resulting loss in genetic diversity and increase in vulnerability to disease, hunting, and random population changes (Soule and Wilcox, 1980).
- Over-harvesting, the taking of individuals at a higher rate than can be sustained by the natural reproductive capacity of the population being harvested; when species are protected by law, harvesting is called "poaching."
- Climatic change, often related to changing regional vegetation patterns; involves such factors as global carbon dioxide build-up, regional effects, such as "El nino" and monsoon systems, and local effects, often involving fire management.

---

from exploitation are often prohibitive for government "owners" which usually lack sufficient resources and local knowledge of management needs to control over-exploitation through the mechanism of enforcing regulations or other restrictions.

Fifth, the discount rates applied by current economic planning tend to encourage depletion of biological resources rather than conservation. While conservation seeks optimum current benefits and broadly equal access to the same stock of resources for future generations, economic analysis usually discounts future benefits and costs because society tends to value benefits sooner rather than later, to consider future costs as being of less significance than costs today, and to assign value to capital in terms of its opportunity cost in the national economy. The higher the discount rate, the greater the likelihood that a biological resource will be mined. Clark (1976) has shown that when discount rates are high and biological growth rates are low (as in whales or tropical forests), the economically efficient use of a resource may be to deplete it, even to extinction; economic activity would be devoted entirely to the interests of the present generation, at the expense of future generations. Further, the higher the discount rate the lower the priority that the planning process will give to investments in conservation (Perrings, *et al.*, 1988); very simply, the returns from such investments may sometimes be so distant in the future that, when discounted, they add little by way of current net benefit. However, the level of the discount rate is a two-edged sword; a low discount rate may make the future better off than the present, but the gain to the future may be in the form of either greater biological diversity or greater consumption (Barrett, 1988).

The selection of a discount rate obviously involves ethical considerations, including such issues as intergenerational equity and the social rate of time preference. It may be relevant to observe that most people work to make the world a better place for their children and grandchildren, often applying a negative discount rate in their personal decisions.

And finally, as Warford (1987b) has observed, conventional measures of national income (such as per capita GNP) "do not recognize the drawing down of the stock of natural capital, and instead consider the depletion of resources, i.e., the loss of wealth, as net income." Many of the national economies of the tropics are based on biological resources, especially forests, which are being depleted at a rate faster than the net formation of capital; as a result, the total assets of the economy are declining even if per capita GNP is growing. Warford estimates that the economic costs of unsustainable forest depletion in major tropical hardwood exporting countries ranges between 4 and 6 percent of GNP, offsetting any economic growth that may otherwise have been achieved. Growth built on resource depletion is clearly very different from that obtained from productive efforts, and may be quite unsustainable.

These rather formidable economic obstacles to the conservation of biological diversity need to be overcome by a series of policy interventions at international, national, and local levels. An essential first step in this process is to determine, or at least estimate, the economic value of biological resources.

APPROACHES FOR DETERMINING THE VALUE OF BIOLOGICAL RESOURCES

Economists have devised a variety of methods for assigning values to natural biological resources (see Barrett, 1988; Brown and Goldstein, 1984; Cooper, 1981; Hufschmidt, *et al.*, 1983; Johansson, 1987; Krutilla and Fisher, 1975; Peterson and Randall, 1984; and Sinden and Worrell, 1979 for details). This multiplicity of ways and means for assessing values is to be expected, because the benefits derived from a biological resource may be measured for one purpose by methods that may not be appropriate for other objectives, and the ways to measure one resource may not be the same for others. The value of a forest in terms of logs, for example, would be measured in quite a different way from the value of the forest for recreation or watershed protection.

Therefore, in order for governments to base decisions on allocating scarce resources on the basis of the best available information, a number of different methods are required to quantify the magnitude and value of the positive and negative impacts. Governments should be seeking means of determining *total* valuation, which require using a wide range of assessment methods. The major approaches are summarized in Box 4, and discussed below.

Direct Values of Biological Resources

Direct values are concerned with the enjoyment or satisfaction received directly by consumers of biological resources. They can be relatively easily observed and measured, often by assigning prices to them. The direct values usually involve consuming the biological resources in question, so they have the potential for stimulating over-exploitation (for the reasons explained above).

Consumptive Use Value. This is the value placed on nature's products that are consumed directly, without passing through a market. When direct consumption involves recreation, as in sport fishing and hunting, most economists estimate consumptive use value

---

Box 4. Classification of Values of Biological Resources

Direct Values
    Consumptive Use Value
    Productive Use Value

Indirect Values
    Non-consumptive Use Value
    Option Value
    Existence Value

---

as the value of the whole recreational experience. The market value of a 5-kilogram salmon, for example, may represent only a fraction of the value an individual places on the experience of catching the fish. These values can be considerable; for example, some 84 percent of the Canadian population participates in wildlife-related recreational activities in a given year, providing them with benefits that they declare to be worth $800 million annually (Fillon, Jacquemot, and Reid, 1985).

While relatively few detailed studies have been carried out on the consumptive use value of species in developing countries, the available information has been well summarized by Myers (1983), Oldfield (1984), Krutilla and Fisher (1975), and Fitter (1986). Of particular interest is the study by Prance *et al.* (1987), which presented quantitative data on the use of trees by four indigenous Amazonian Indian groups. "Use" was defined rather narrowly, including as food, construction material, raw material for other technology, medicinals, trade goods, and other; uses as firewood or as food for harvested animals were not included. The percentage of tree species used by the various groups varied from 48.6 to 78.7,

indicating that the rainforests of Amazonia contain an exception-
ally large number of species that are useful to local people.

Consumptive use values seldom appear in national income accounts, but no serious obstacles appear to prevent the inclusion
of at least some consumptive use values in such measures as Gross
Domestic Product (GDP). For example, firewood and dung are used
to provide over 90 percent of the total primary energy needs in
Nepal, Tanzania, and Malawi and exceed 80 percent in many other
countries (Pearce, 1987a).

In Africa, harvested species make a considerable contribution
to human welfare in the form of food for rural people, and especially to the poorest villagers living in the most remote areas. Much
of this is consumed directly rather than being sold in the marketplace, but the value is nonetheless significant and economic values
can be assigned. In Botswana, over 50 species of wild animals, ranging from elephants to rodents, bats and small birds provide animal protein exceeding 90 kg per person per annum in some areas
(some 40 percent of their diet); over 3 million kg of meat is obtained
yearly from springhare alone. In Ghana, about 75 percent of the
population depends largely on traditional sources of protein supply, mainly wildlife, including fish, insects, caterpillars, maggots and
snails. In Nigeria, game constitutes about 20 percent of the mean
annual consumption of animal protein by people in rural areas (including 100,000 tons of the giant rats known as "grasscutters"—
Myers, 1988), while 75 percent of the animal protein consumed in
Zaire comes from wild sources. Senegal's population of 5 million
consumes at least 373,631 metric tons of wild mammals and birds
per year (Sale, 1981).

Consumptive use value can be assigned a price through such
mechanisms as estimating market value if the product were sold
on the market instead of being consumed. In Sarawak, Malaysia,
for example, a detailed field study found that wild pigs harvested
by hunters had a market value of some $100 million per year (Cal-
decott, 1988).

Productive Use Value. This value is assigned to products which
are commercially harvested, and is therefore often the only value
of biological resources which is reflected in national income accounts. Estimates of such values are usually made at the production end (landed value, harvest value, farmgate value, etc.) rather
than at the retail end, where values are much higher; for example,

the estimated production value of cascara in the US is $1 million per year, but the retail value is $75 million per year (Prescott-Allen and Prescott-Allen, 1986).

Productive use of such biological resource products as fuelwood, timber, fish, animal skins, musk, ivory, medicinal plants, honey, beeswax, fibres, gums, resins, rattans, construction materials, ornamentals, animals harvested for game meat, fodder, mushrooms, fruits, dyes, and so forth can have a major impact on national economies. In addition, wild biological resources contribute to the production of domesticated resources in several ways:

- wild species serve as sources of new domesticates;
- wild genetic resources are used to improve established domesticates (a contribution valued in the tens of billions of dollars per year);
- rangeland and wild forage species contribute to livestock production;
- wild pollinators are essential to many crops; and
- wild enemies of pests help control their depredations on crops.

According to Prescott-Allen (1986), the productive use value of wild genetic resources demonstrates that genetic resources are indispensable to modern agriculture, that most of them come from a country other than where they are utilized, that the turnover of domestic genetic resources is rapid, and that use of new genetic resources is increasing (therefore requiring the lines of supply from other countries to be kept open and a great diversty of genetic resources to be maintained).

Productive use value can be derived directly from the market demand curve for the resources consumed. The demand curve is a schedule of consumers' willingness to pay for various quantities of the resource. Where close substitutes are available, the demand curve will be fairly flat and the productive use value can be approximated by market price. Where close substitutes are not available, there exists a "consumers' surplus" over and above the market price. In this case, use of price data may severely underestimate productive use value.

Prescott-Allen and Prescott-Allen (1986), in a major path-breaking study which demonstrated how the dollar value of biological resources can be estimated, carried out a detailed analysis of the contribution wild species of plants and animals made to the American economy, concluding that some 4.5 percent of gross domestic product (GDP) is attributable to wild species. The combined contribution to GDP of wild harvested resources averaged some $87 billion per year over the period 1976 to 1980.

The contribution of wild species and ecosystems to the economies of developing countries is usually far greater (in percentage GDP terms) than it is for an industrialized country like the USA. Timber from wild forests, for example, is the second leading foreign exchange earner for Indonesia (after petroleum), and throughout the humid tropics governments have based their economies on the harvest of wild trees; total exports of wood products from Asia, Africa, and South America averaged $8.1 billion per year between 1981 and 1983 (WRI/IIED, 1986).

Non-wood forest products can also be of considerable value. Indonesia, for example, earned some $200 million in foreign exchange from non-wood forest products in 1982 (Gillis, 1986), while non-wood forest products in a recent year provided 40 percent of the total net revenues accruing to the Indian government from the forestry sector, and 63 percent of the forestry exports (Gupta and Guleria, 1982). In comparing wood and non-wood forest resources, Myers (1988) concludes that a tropical forest tract of 500 square kilometers could, with effective management, "produce a self-renewing crop of wildlife with a potential value of at least $10 million per year, or slightly more than $200 per hectare. These revenues contrast with the return from commercial logging in the area of only a little over $150 per hectare. Moveover, with present timber-harvesting practices, commercial logging tends to be an ecologically disruptive procedure, whereas wildlife harvesting can leave forest ecosystems virtually undisturbed" (see also Case Study 18 for an example from Brazil).

The returns from wildlife will be far less in drier habitats, though often exceeding alternative uses. In Zimbabwe's Zambezi Valley, for example, Cumming (1985) estimates that potential gross returns from wildlife utilizations amount to $12 per hectare. "These returns," he states, "are as good if not better than returns from the best-run commerical beef ranches in the country and the profit margins are probably higher."

However, as will be demonstrated by the discussion of indirect values below, the market price is not always an accurate representation of the true economic value of the resource, and does not deal effectively with questions of distribution and equity. It is also apparent that consumers may value resources in ways different from producers; California redwoods are valued by consumers of scenic beauty differently than by consumers of lumber products, but no market is available to mediate these claims.

Species harvested for use in making other products such as drugs also have productive use value, with their price often being derived from the value of the final product. In the OECD countries, for example, the retail market value of both prescription and over-the-counter plant-based drugs is estimated to have been about $43 billion in 1985. When social benefits from increased good health—wages not lost, health care costs averted, the value individuals place on better health, etc.—are included, it is estimated that the pharmaceutical economic value of plant-based drugs ranges from $200 billion to $1.8 trillion annually for all OECD countries (Principe, 1988a); but recall that this is retail value, not productive use value. In developing countries, where medicinal plants are even more important elements in health care, the contribution is likely to be far greater, in percentage terms; though reliable statistics are difficult to come by, Prescott-Allen and Prescott-Allen (1982), Myers (1983) and Oldfield (1984) provide useful summaries.

Indirect Values of Biological Resources

Indirect values, which deal primarily with the functions of ecosystems (here called "environmental resources"), do not normally appear in national accounting systems but they may far outweigh direct values when they are computed. These values tend to reflect the value of biological diversity to society at large rather than to individuals or corporate entities.

Direct values often derive from indirect values, because harvested species of plants and animals are supported by the goods and services provided by their environments. Species without consumptive or productive use values may play important roles in the ecosystem, supporting species that are valued for their productive or consumptive use. In Sabah, for example, recent studies suggest that high densities of wild birds in commercial Albizia plantations limit the abundance of caterpillars that would otherwise defoliate the trees; the birds require natural forest for nesting.

Nonconsumptive Use Value. Environmental resources—generally speaking, nature's functions or services rather than goods—provide value without being consumed, traded in the market-place, or reflected in national income accounts. Still, efforts are being developed to evaluate economically the benefits provided by these resources (Oldfield, 1984, Peterson and Randall, 1984; Sinden and Worrell, 1979; de Groot, 1986). It is apparent that the benefits of environmental services are much easier to measure at the local level than at the global level; quantifying the hydrological benefits of a watershed, for example, is relatively straight-forward, while measuring the value of the global carbon cycle would be a daunting exercise.

A nonconsumptive use such as organized tourism based on biological resources (such as visits to a national park) can often provide a powerful economic justification for conserving biological resources, particularly when protected areas are a primary attraction for visitors to a country. In Kenya, for example, tourism is the leading foreign exchange earner, and much of the tourism is based on Kenya's system of protected areas. Each lion in Amboseli National Park has been estimated to be worth $27,000 per year in visitor attraction, and each elephant herd is worth $610,000 per year; the park yields net earnings—mostly from tourism—of about $40 per hectare per year, some 50 times the net profit under the most optimistic agricultural projection (Western, 1984). (As will be discussed in Chapter 4, the challenge comes in bringing the economic benefits of tourism to the local community which is paying the opportunity cost of not harvesting elephants.)

Species can also have nonconsumptive use value, as in bird watching and some scientific research (especially ecological field studies). People also derive indirect nonconsumptive use value from species through media such as film, video, and literature.

Many nonconsumptive values have considerable economic impact. Oldfield (1984) reports that in Massachusetts, a study of wetlands estimated the capitalized value (at 5.375 percent) at $147,900 per hectare for wetlands with a high capacity for provision of water supply, flood control, wildlife, and recreational and esthetic benefits. The value of coastal marshes, which provided primary productivity which in turn supported offshore commercial and recreational fishing industries, was determined to be $4,938/ha/year.

The U.S. National Marine Fisheries Service estimates that the destruction of U.S. coastal estuaries between 1954 and 1978 cost the nation over $200 million annually in revenues lost from commercial

Box 5. Non-Consumptive Benefits of Conserving Biological Resources

The benefits accruing to society in return for investments in conserving biological resources will vary considerably from area to area and resource to resource. Most such benefits will fall into one or another of the following categories:

- Photosynthetic fixation of solar energy, transferring this energy through green plants into natural food chains, and thereby providing the support system for species which are harvested;
- Ecosystem functions involving reproduction, including pollination, gene flow, cross-fertilization; maintenance of environmental forces and species that influence the acquisition of useful genetic traits in economic species; and maintenance of evolutionary processes, leading to constant dynamic tension among competitors in ecosytems;
- Maintaining water cycles, including recharging groundwater, protecting watersheds, and buffering extreme water conditions (such as flood and drought);
- Regulation of climate, at both macro- and micro-climatic levels (including influences on temperature, precipitation, and air turbulence);
- Production of soil and protection of soil from erosion, including protecting coastlines from erosion by the sea;
- Storage and cycling of essential nutrients, e.g., carbon, nitrogen, and oxygen; and maintenance of the oxygen-carbon dioxide balance;
- Absorption and breakdown of pollutants, including the decomposition of organic wastes, pesticides, and air and water pollutants; and
- Provision of recreational-esthetic, sociocultural, scientific, educational, spiritual, and historical values of natural environments.

and sport fisheries. Another estimate placed the economic value of a hectare of Atlantic Spartina marsh at over $72,000 a year. According to the U.S. Army Corps of Engineers, retaining a wetlands complex outside of Boston, Massachusetts realized an annual cost savings of $17 million in flood protection alone (this figure did not include the many other benefits—such as sediment reduction, fish and wildlife production, and esthetic values—that the wetlands afforded area residents) (Hair, 1988).

Option Value. The future is uncertain, and extinction is forever. Prescott-Allen and Prescott Allen (1986) suggest that society "should prepare for unpredictable events, both biological and socio-economic. The best preparation in the context of wildlife use is to have a safety net of diversity—maintaining as many gene pools as possible, particularly within those wild species that are economically significant or are likely to be." Option value is a means of assigning a value to risk aversion in the face of uncertainty.

Nobody can determine today which species will be most valuable tomorrow, or how much genetic diversity in wild relatives of domestic plants will be necessary for supporting agriculture. One outstanding illustration of the possible magnitude involved was the discovery in 1979 of a new species of maize (called teosinte by the local people) on a small hillside in Mexico, which was in the midst of being cleared; the species was remarkable in being a perennial grass rather than an annual like other types of maize. Hanemann and Fisher (1985) have shown that under certain assumptions, teosinte may prove to have a value of $6.82 billion annually for its contribution to creating a perennial hybrid of corn (maize).

Protected areas preserve a reservoir of continually evolving genetic material—irrespective of whether the values of that material have yet been recognized—which enables the various species to adapt to changing conditions. The plants and animals conserved may spread into surrounding areas where they may be able to be cropped at some future date, or may eventually contribute genetic material to domestic crops or livestock. Protected areas can therefore be seen as a means for nations, especially those in the species-rich tropics, to keep at least part of their biological resources intact for the future benefit of their populace.

Therefore, society as a whole may be willing to pay to retain the option of having future access to a given species or level of diversity. As the demand for biological resources grows while the supply continues to dwindle (if current trends continue), their value is likely to increase. Therefore, some economists suggest that conventional cost-benefit relationships need to incorporate mechanisms to deal with the probability of higher future values and the irretrievability of lost opportunities to preserve natural environments and genetic material.

Of less direct relevance to survival issues, but still important at the international level, is the option value of protected areas for tourism. Some people may not know whether they will visit a protected area in the future, but attach a value to having the option of doing so. Cicchetti and Freeman (1971) even claim that the option to visit wild areas containing significant biological diversity has values (hope, opportunity, dreams, fellowship, satisfaction, etc.) that are independent of the values of actually going.

The "total use value" for a biological resource is given by an individual's maximum willingness to pay for a project which preserves his or her option to make use of the good or service in the future. This total use value—the sum of the option value and the expected value of actually making use of the good or service (the expected consumer's surplus)—is called "option price." It is generally agreed that option price is the most appropriate measure of use value (Graham, 1981). Applying total use value involves difficult equity issues, because wealthier individuals can obtain a large "vote" with a small portion of their income, while less wealthy individuals may have no disposable income with which to "vote."

Ultimately, the determination of option value is an empirical question which needs to be answered with the specifics of each case.

It is apparent that few policies can fully guarantee the long-term conservation of a given biological resource, but sound policies increase the probability that conservation will be successful. "Access option value" involves willingness to pay for an increase in the probability of gaining future access to the resource (Gallagher and Smith, 1985).

Another form of option value has been called "serendipity value" (Pearsall, 1984), the potential that each species—especially those that have not yet been discovered, or their characteristics fully explored—may be found to have for human use as food, genetic material, medicine, or other raw material. The popular conservation literature assigns great importance to this serendipity value (see, for example, Myers 1984 and Schultes and Swain, 1976). The continuing flow of new discoveries from natural ecosystems—such as the role of plants in fixing heavy metals from the soil (Baker, Brooks, and Reeves, 1988) and the role of animals as indicators of ecosystem responses to air pollution (Newman and Schereiber,

1984)—indicate that serendipity value is significantly greater than zero. Further, new breakthroughs in biotechnology suggest that biological diversity may have even greater value in the future than it does at present.

It can often be shown that a given development project will cause the irreversible destruction of some biological resources. An option would be to postpone the development project until the value of these resources is known. Uncertainty about the value of biological diversity will not be resolved by clear-cutting the forest or by constructing a hydroelectric dam, but these projects can still be undertaken at a later date. The value of being able to learn about future benefits that would be precluded by the project—"quasi option value"—is positive provided the information is solely time-dependent (see Fisher and Hanemann, 1987).

Existence Value. Many people, especially in the industrial nations, also attach value to the existence of a species or habitat that they have no intention of ever visiting or using; they might hope that their descendants (or future generations in general) may derive some benefit from the existence of these species, or may just find satisfaction that the oceans hold whales, the Himalayas have snow leopards, and the Serengeti has antelope. The ethical dimension is therefore important in determining "existence value," which reflects the sympathy, responsibility, and concern that some people may feel toward species and ecosystems. An accurate cost-benefit analysis of such values is clearly impossible, but the magnitude of these values is suggested by the sizeable voluntary contributions to private conservation agencies in the developed world by people who do not expect to visit or use the resource they are helping to conserve (WWF alone receives nearly $100 million per year in such donations). Existence price is similar to option price, but is based on the perception that no probability of use exists.

A particularly important variant of existence value may be called "bequest value," the vicarious benefit received now because someone who may not yet exist may benefit in some unidentified way from the future existence of some biological resource. Bequest value is often considered to provide much of the economic justification for preserving natural lands (Krutilla and Fisher, 1975), and seems to dominate all other benefits of wilderness in the minds of some people (Pearsall, 1984). It is one of the best means of dealing with problems of inter-generational equity.

THE BENEFITS OF PROTECTING HABITATS

One of the best-known and most effective ways of conserving many biological resources is through establishing legal regimes which provide protection to both the habitat and the biological diversity contained by that habitat. Virtually all countries today contain protected areas, and the area protected currently totals nearly 4 percent of the world's land surface; Bhutan, Botswana, Chile, Malawi, New Zealand, Rwanda, Senegal, Sri Lanka, and Togo have each established protective regimes which exceed 10 percent of their territory (though effectiveness of protection varies considerably), a reasonable minimum standard suggested by IUCN (1984) and the World Commission on Environment and Development (1987).

Based on a detailed review of the world's protected areas, MacKinnon *et al.* (1986) determined that protected areas have at least the main benefits cited below, all of which have economic values which can be estimated in various ways in each specific case. These benefits cover the entire gamut of direct and indirect values, though most of them are non-consumptive. The case studies illustrate how various protected areas have recognized these values, and promoted them through the use of incentives.

1) *Stabilizing hydrological functions.* Natural vegetation cover on water catchments in the tropics regulates and stabilizes water run-off. Deep penetration by tree roots or other vegetation makes the soil more permeable to rainwater so that run-off is slower and more uniform than on cleared land. As a consequence, streams in forested regions continue to flow in dry weather and floods are minimized in rainy weather. Daniel and Kulasingham (1974) showed that in Malaysia, the peak runoff per unit area of forested catchments is about half that of rubber and oilpalm plantations, while the low flows are roughly double. In some cases these hydrological functions can be of enormous value. For example, Venezuela's Canaima National Park safeguards a catchment feeding hydroelectric developments which are so important that the government recently tripled the size of the park to 3 million ha to enhance its utility for watershed protection (Garcia, 1984). (See also Case Study 13 for an example from Honduras.)

2) *Protecting soils.* Exposed tropical soils degrade quickly due to leaching of nutrients, burning of humus, laterization of

minerals and accelerated erosion of topsoil. Good soil protection by natural vegetation cover and litter (especially significant in grassland ecosystems) can preserve the productive capacity of the reserve itself, prevent dangerous landslides, safeguard coastlines and riverbanks, and prevent the destruction of coral reefs and freshwater and coastal fisheries by siltation. A startling example of soil conservation is provided by Nepal's Royal Chitwan National Park, where villagers have cleared and grazed the north bank of the Rapti River (which forms the park boundary) so intensively that it has been the subject of rapid erosion. On the south bank, within the park, the protected vegetation binds the soil so that when monsoon rains swell the Rapti it is the north bank that is washed away. As a result, the course of the river has shifted and in less than a decade roughly one square kilometer has been taken from villagers and added to the park by natural forces (Roberts and Johnson, 1985). Myers (1988) quotes evidence that in Malaysia, erosion from maize croplands and oilpalm plantations can be 11 times higher than from primary rain forest, from peanut plantations 12 times higher, from tea plantations 20 or more times higher, from vegetable croplands 34 times higher, and from bare soil 45 times higher.

3) *Contributing to stability of climate.* Growing evidence suggests that undisturbed forest helps to maintain the rainfall in its immediate vicinity by recycling water vapor at a steady rate back into the atmosphere and by the canopy's effect in promoting atmospheric turbulence. This may be particularly important in the production of dry season showers which are often more critical for settled agriculture than the heavier monsoon rains (Dickinson, 1981; Henderson-Sellers, 1981). Forest cover also helps to keep down local ambient temperatures, benefitting surrounding areas both for agriculture (lowered transpiration levels and water stress) and for human comfort.

4) *Conserving renewable harvestable resources.* While intensively managed forest plantations of carefully selected species will almost always out-produce natural forest stands in terms of biomass production, the combined economic benefits (including wood and non-wood products, and externalities) of natural forests often surpass that of plantations. The quantity

and value of natural materials that can be harvested on a sustainable basis will vary considerably, depending upon the protection category of the reserve, and may be of as much value to the local community as any alternative land-use. In Nepal's Royal Chitwan National Park, for example, the local people are allowed into the park during a specific two-week period each year to harvest thatch grass, worth some $1 million per year to the 59,000 local villagers involved in the harvest. Since the area around Chitwan has been denuded of natural vegetation, the park now provides virtually the only source of thatch, the most important traditional roofing material in the region (Mishra, 1984).

5) *Protecting genetic resources.* People are known to make use of some 15,000 species of wild plants and animals for foods, medicines and utilities, many to a commercially important degree. Several thousand more species may be of potential use (serendipity value). All domestic plants and animals were originally derived from the wild and many can only be maintained and improved by regular recrossing with wild forms and relatives. The short- and long-term values of these genetic resources are enormous and most improvements in tropical agriculture and silviculture depend on their preservation. Moreover, the gene pool value of reserves will increase as remaining natural habitats become more scarce. Protected areas are therefore of great value as *in situ* genebanks but only as long as they are protected.

6) *Preserving breeding stocks, population reservoirs and biological diversity.* Reserves may protect crucial life stages or elements of wildlife populations that are widely and profitably harvested outside reserves. They are sources of seed dispersal, wildlife, and fish spawning areas, often providing considerable economic returns; in India, for example, a partially protected mangrove forest produced some 110 kg of prawns/ha/year, while a similar unprotected mangrove produced just 20 kg/ha/year. Protected areas also act as "refugia" wherein biological diversity can be maintained, and this is often one of their strongest justifications (especially in the context of this paper).

7) *Maintaining the natural balance of the environment.* The existence of a protected area may help maintain a more natural balance

of the ecosystem over a much wider area. Protected areas afford sanctuary to breeding populations of birds which control insect and mammal pests in agricultural areas. Bats, birds and bees which nest, roost, and breed in reserves may range far outside their boundaries and pollinate fruit trees in the surrounding areas. Ledec and Goodland (1986) have shown how the production of Brazil nuts depends on a variety of poorly-known forest plants and animals. Male euglossine bees which pollinate the flowers of the Brazil nut tree gather certain organic compounds from epiphytic orchids to attract females for mating. The hard shell covering the nut is opened naturally only by the forest-dwelling agouti (a large rodent), thereby enabling the tree to disperse. Thus, maintaining Brazil nut production appears to require conserving enough natural forest to protect bee nesting habitat, other bee food plants, certain orchids and the trees upon which they grow, the insects or hummingbirds that pollinate the orchids (and all their necessities in turn), and agoutis. Another good example comes from Tanzania, where the poaching and uncontrolled hunting of elephants to the south-east of Tarangire National Park led to bush encroachment which caused an increase in tsetse flies which in turn led to a livestock reduction in the area; conservation of elephants would have enhanced the productivity of the livestock industry.

8) *Supporting tourism and recreation.* At the national level, tourism frequently brings in valuable foreign exchange and at the local level stimulates profitable domestic industries—hotels, restaurants, transport systems, souvenirs and handicrafts, and guide services. Returns on tourism to natural areas are often considerable; the Virgin Islands National Park, for example, earned an estimated 10-fold annual return in benefits over investments (Island Resources Foundation, 1981). Annual cash income from tourism to marine parks in the Caribbean include such figures as $2 million for Caroni Swamp (Trinidad), $5 million for Bonaire Marine Park, $14 million for British Virgin Islands parks, and $50 million for Cayman Islands protected areas (Heyman, 1988). In some societies, local communities as well as other domestic visitors benefit from the recreational facilities provided by most categories of protected areas. Benefits from recreation and tourism in protected areas

are likely to become ever more valuable as the availability of other wild areas is further reduced.

9) *Creating employment opportunities.* Apart from the employment created within the protected area itself, additional employment is generated by auxiliary services, tourist development, road improvements and professional services. Such benefits are particularly relevant where the land allocated for the protected area has little or no value for agriculture. In the arid Kirthar Mountains of Pakistan, for example, the 108 local villagers who are employed by the Kirthar National Park have provided an economic boost to the region; and in Nepal, one nature tourism company alone directly employs some 5,000 people.

10) *Providing facilities for research, education, and monitoring.* People have much to learn on the subject of how to get better use from biological resources. Tropical agriculture in particular is often based on plants which are disease-and pest-prone, require fertilizers and result in soil degradation. Much applied research still needs to be done in natural tropical ecosystems to find the secrets of high, stable productivity on poor soils. Protected areas provide excellent living laboratories for such studies, for comparison with other areas under different systems of land use and for valuable research into ecology and evolution. Unaltered habitats are often essential for certain research approaches, providing controls against which the changes brought about by other forms of land use may be measured and assessed. Protected areas provide valuable sites for school classes and university students to gain practical education in the fields of biology, ecology, geology, geography and socio-economics. Such uses can extend to, and ultimately benefit, a large proportion of the local population. Research can also bring direct benefits; for example, Costa Rica's Guanacaste National Park currently brings some $200,000 per year into the local economy from research projects, through purchase of supplies, hire of local assistants, and other contributions to the economy (Allen, 1988).

Costs of Replacing Biological Resources

In many cases, biological resources can be easily transformed into revenue through harvesting. However, an ecosystem which has been depleted of its economically-important species or a habitat which has been altered to another use cannot be re-built out of income the way industrial capital can be. The costs of re-establishing forests, or reversing the processes of desertification, can be extremely high and far exceed any economic benefits from over-harvesting the biological resources the ecosystem contains.

Therefore, when discussing benefits of conserving biological resources it is also useful to consider the environmental costs of depletion, in terms of the time and effort required to restore the resource to its former productivity. Economists often use the concept of "replacement value" to estimate what would be required to recover a biological resource which has been lost or to restore a depleted resource to its former productivity. (Note that species once lost cannot be recovered, and that some degradation processes—such as desertification—may be essentially irreversible in time scales relevant to modern society.) The costs of reforestation, for example, indicate the value that governments are willing to spend to maintain the original forest. Natural wetlands which serve to purify water may need to be replaced by water purification plants if they are destroyed. For example, replacing the tertiary waste treatment services provided by marshes in Massachusetts has been calculated at about $123,000/ha, and for removal of phosphorus alone, $47,000/ha (Oldfield, 1984). The cost of restoring a wetland or rehabilitating a marsh could be even greater.

Replacing the efficiency lost to dams by erosion can be assigned a value, to the extent that the siltation is caused by erosion related to upstream depletion of forests. Pearce (1987b) concludes that siltation of reservoirs feeding hydropower facilities involves a loss of some 148,000 gigawatt hours which, at US$15 per barrel, would cost some $4 billion per annum to replace using thermal generation. "Such calculations are crude," says Pearce, "and they need considerable refinement to try to assess the contribution of erosion due to resource degradation. But a cost in terms of half the maximum rate would be $2 billion per annum and this only in terms of replacement value."

Costs of Protecting Biological Resources

Protecting biological resources also has costs, particularly for the local people living around protected areas. Retaining natural vegetation implies an opportunity cost; a tropical forest might have instead been logged and converted into a plantation of trees or sugar cane, or a conserved mangrove forest could instead have been converted into fish ponds (though a thorough economic analysis might well demonstrate that conversion of mangroves to fish ponds usually results in a reduction in productivity and a decline in the well-being of local people). Protection may prevent people from exploiting resources as freely as they might wish, and they may thereby feel deprived of a resource to which they believe they deserve free access. Further, conserving wildlife inevitably means conflicts between predators and domestic stock, loss of field crops to large grazing mammals, and other perceived threats to human welfare.

This descriptive list of values and costs of protecting biological resources provides a basis for determining the total value of any protected area or other system of biological resources, indicating that a range of approaches to valuation are required. Not all of the benefits are derived from every reserve or resource system, and many of them need some investment if they are to contribute to the economy; similarly, subsidies can overcome the costs to local people of establishing a protected area and cash awards can compensate villagers for the cost of losing part of a crop to a wild mammal or bird. But the value of conserving biological resources tends to be cumulative, and the total value to the region as a whole can be considerable.

THE CONCEPT OF MARGINAL OPPORTUNITY COST

Any allocation of land involves choices. For example, whether establishing a reserve is the best land use for a particular area will depend on the total of these costs and benefits compared with the total potential costs and benefits that would have been attainable if the area were designated for conversion into a use which conserved a lower level of biological diversity. Conserving the wrong areas can be expensive.

Obviously, it is easier to find socio-economic justification for conserving biological resources in marginal lands than in areas of high agricultural or urban potential, all else being equal. But careful analysis can often demonstrate the value of a program to conserve biological diversity, to both the national economy and the nearby communities.

For example, the South Australian Department of Environment and Planning found that the total cost to the community of the national parks on Kangaroo Island was $500,000 per year. These costs included park establishment and management costs, costs of parks to neighbors (productivity losses, fencing repairs, native animal culling, and fire control), and the opportunity cost of the parks. The quantifiable benefits of the parks were estimated at between $4.3 million and $5.6 million, including visitor expenditure associated with parks, expenditure by tourism operators and other flow-on effects. Other benefits, which were not quantified, included local community recreation benefits, education, research, and conservation (Lothian, 1985).

In seeking to promote a more integrated approach to assessing the value of biological resources, economists have developed the concept of "Marginal Opportunity Cost" (MOC). MOC is the true cost borne by society for an action or policy which depletes a biological resource; in a system of taxing destructive resource users, the tax that users pay for activities which deplete biological resources would ideally be equal to MOC (Warford, 1987a; Pearce, 1987c). "Opportunity cost" refers to the value corresponding to the best alternative use to which a particular biological resource could be put if it were not being used for the purpose which is being costed. If, for example, the highest-valued alternative use of the forest is as a national park, then the opportunity cost of logging the forest is the value of the national park; the full cost of logging should include this cost plus the opportunity cost of the labor and equipment used in the logging operations. If the benefits of logging (price times output) do not exceed the social opportunity costs, then logging should not take place. Clearly, the opportunity cost of logging must take into account all the values discussed above.

MOC has three elements: the direct and indirect cost to the user of depleting a biological resource; the benefits forgone by those who might have used the resource in the future (the option value); and the costs imposed on others (external costs). While the proportion of MOC that can be assigned a monetary value will vary from case to case, it still provides a useful analytical framework within which the various methods of determining value of biological resources can contribute to decision making. In the case of establishing a national park, MOC may indicate the amount of money required to compensate those who have forgone the income that would have come from logging the forest, if indeed that income is more than the net benefits of conserving the forest.

As Pearce (1987a) points out, biological resources can be used sustainably at a range of stock levels; MOC may indicate to a government that it should reduce permanently its stock level of a given biological resource to generate income to support development activities. The choice of the optimal level of stock is determined by comparing MOC with the marginal benefits at different stock levels, assuming that the managed stock is not being harvested at a rate faster than regeneration.

The obstacles to the wide use of MOC have involved: different viewpoints over what constitutes reasonable expectations for costs, revenues, and the choice of discount rate; false claims based on development intentions which may not be genuine; and insufficient funding for the compensating authorities (CEMP, 1988).

Still, MOC is a very useful tool for making decisions about allocation of resources. It serves "to focus attention on the relationship between acts of resource depletion now and their effects elsewhere in the economy and in the future. Moreover, it is linked to a view of the development process which emphasizes the role of renewable natural resources, and which argues that development and environmental preservation are inseparable parts of the process of social improvement" (Pearce, 1987b). It can be used as a means by which those who will lose from having restrictions placed on their use of biological resources can be compensated to recover the value of their lost opportunity. MOC can be expressed either in terms of the annual net revenue forgone, or the difference between the land value in restricted and unrestricted use.

CONCLUSIONS

Decision makers need to know the manner in which people benefit and lose as a result of conserving biological resources, as well as the values and costs of the goods and services that create the gains and losses. This requires reasonably detailed knowledge of the status and trends of biological resources, adoption of methodologies for determining the values of these resources, and continuous programs of research and monitoring. Governments also need to be clear and explicit about national objectives for conserving biological resources.

As human pressure on land increases, it becomes more important to put an economic value on both the direct and indirect benefits provided by biological resources and to predict the likely immediate and future costs to the community if the diversity of these resources is depleted.

Many biological resources are seriously under-priced. For example, the current market price of timber does not reflect either the external costs or the unsustainable nature of its production. Raising the price of tropical timber to reflect its true resource cost would greatly reduce demand, thereby reducing the need for protected areas. It would also require governments to seek alternative sources of foreign exchange (which will in any case be required when forests are depleted).

While it is possible to justify conserving biological resources through identifying qualitative benefits, even partial valuation in monetary terms of the benefits of conserving biological resources is helpful in providing at least a lower limit to the full range of benefits. Various methodologies exist for estimating the monetary value of physically quantified biological resources (World Bank, 1986).

Further, as Warford (1987b) states, "If economic methods are to be successful, it is crucial that their limitations be understood and continually kept in mind. In particular, it should be recognized that value judgments about distributional and irreversible effects are unavoidable, but quantification in monetary terms of as many variables as possible is important in crystallizing those issues involving implicit value judgements which may otherwise be ignored."

Ehrenfeld (1988) carries this warning a step further: "It is certain that if we persist in this crusade to determine value where value ought to be evident, we will be left with nothing but our greed when the dust finally settles. I should make it clear that I am referring not just to the effort to put an actual price on biological diversity but also to the attempt to rephrase the price in terms of a nebulous survival value. . . . As shown by the example of the faltering search for new drugs in the tropics, economic criteria of value are shifting, fluid, and utterly opportunistic in their practical application. This is the opposite of the value system needed to conserve biological diversity over the course of decades and centuries."

Further, many scientists will argue, nobody knows enough about any gene, species, or ecosystem to be able to calculate its ecological and economic worth in the larger scheme of things. And, Ehrenfeld (1988) adds, "The species whose members are the fewest in number, the rarest, the most narrowly distributed—in short, the ones most likely to become extinct—are obviously the ones least likely to be missed by the biosphere." On the other hand, many of these may be greatly missed by people; a dramatic example might be the only population of the wild rice *Oryza nivara* that is the only source of resistence to grassy stunt virus.

Such perspectives are well worth bearing in mind, but the fact remains that the major decisions which are affecting the status and trends of biological resources are based on economic factors, including the establishment of their value. Those interested in the effective management of biological resources cannot avoid addressing issues of economic value, even realizing the ethical limitations of these issues.

The mainstream economic approach today, as exemplified by USAID (1987), is to complete a particular form of utilitarian calculation expressed in money values and including (in raw or modified form) the commercial values that are expressed in markets. However, it expands the account to include things that enter human preference structures but are not exchanged in organized markets. This extension and completion of a utilitarian account, where conservation of biological resources is at issue, is useful because it demonstrates that commercial interests do not always prevail over economic arguments (Randall, 1988).

Randall (1988) concludes: "The claim that it is useful to complete this utilitarian account does not depend on any prior claim that the utilitarian framework is itself the preferred ethical system. Environmental goals that may be served by arguments that the biota has rights that should be considered, or that it is the beneficiary of duties and obligations deriving from ethical principles incumbent on humans, may also be served by completing a utilitarian account that demonstrates the value implications of human preferences that extend beyond commercial goods to include biodiversity. Some people would argue that a complete discussion of the value of biodiversity should extend beyond utilitarian concerns. Even these people would, presumably, prefer a reasonably complete and balanced utilitarian analysis to the truncated and distorted utilitarian analysis that emerges from commercial accounts."

Different approaches to valuation are relevant at different levels. At the local level, consumptive use value is often the most relevant, while national governments tend to be most interested in productive use value, often in terms of the foreign exchange earned. While many products from biological resources are traded internationally, the world community is also likely to be interested in existence value and non-consumptive use value. Wealthy individuals or nations may be more concerned about option value than nations which are carrying a heavy debt burden and may be forced into unsustainable productive uses. Development assistance agencies may be particularly interested in replacement value, as they work to rehabilitate degraded ecosystems.

But whatever methodology is used, valuation is only the first step. It informs planners and local people about how important biological diversity may be to national development objectives, and may demonstrate that an area is important for the biological resources it contains.

The second step is to determine how these areas can be conserved. It is here that economic incentives and disincentives can play their important role in ensuring that the benefits suggested above are in fact delivered to the community, and that the community in turn is enabled to protect the resources upon which its continued prosperity depends.

CHAPTER THREE

ECONOMIC INCENTIVES: WHAT THEY ARE AND HOW THEY CAN BE USED TO PROMOTE CONSERVATION OF BIOLOGICAL DIVERSITY

INTRODUCTION

Once a government has determined its objectives for maintaining biological diversity, estimated the goods and services it wishes to receive on a sustainable basis from effectively managed biological resources, and considered the economic costs and benefits of providing these resources, a number of options are available for converting these policies into reality. These include:

- institutional mechanisms, such as establishing and maintaining government agencies for implementing biological diversity objectives (usually in the form of a wildlife management or national park management agency) and agencies for coordinating the various government activities affecting biological diversity (e.g., ministries of environment, national planning agencies);
- implementing research programs by universities and national research agencies;
- enacting and enforcing laws and regulations; and
- developing economic methods for encouraging behavior which conserves biological resources or discouraging behavior which depletes these resources.

The first three mechanisms—i.e., institutions, research, and legislation—are reasonably well known and will be considered only peripherally in this paper (for further information on these mechanisms, see Lyster, 1986; Lausche, 1980; OTA, 1987; Schonewald-Cox, 1983). The last—basically economic incentives and disincentives—is insufficiently applied and will be the focus here.

Insufficient attention is only one of the reasons for focussing on economic incentives, and perhaps not even the most important one. More important is the observation that **current institutions, research, and legislation have failed to conserve the level of biological diversity required for the welfare of society.**

Even the most enlightened governments are having difficulties in protecting their natural diversity in the current economic climate. For example, Costa Rica, with Central America's most effective system of managing biological resources, has seen its economy assailed by inflation, falling commodities prices, and external debt that it has had to reduce the staff of its National Park Service by 20 percent from 1979 to 1988, and the 1988 budget is just 40 percent of the 1979 budget (at constant dollar values); during that same period, the size of Costa Rica's park system more than doubled, and the management challenges for the shrinking staff and budget grew apace (Barborak, 1988a).

Throughout the world, insufficient personnel and budgets mean that biological resources are inadequately protected by government programs, and biodiversity is suffering as a result. New mechanisms for conserving biological resources are clearly required.

This paper suggests that the best strategy for governments today is to supplement their current systems of administration, research, and legislation with comprehensive systems of economic incentives and disincentives for promoting the conservation of what the government has determined to be the optimum levels of biological diversity for the nation. When governments are not able to make a prior determination of the optimum (which will often be the case, especially in the tropics), these incentives at least can move in the general direction deemed appropriate.

THE NATURE OF INCENTIVES AND DISINCENTIVES

Incentives motivate desired behavior, and disincentives discourage behavior which is not desired. For the purposes of this discussion, an incentive is any inducement which is specifically intended to incite or motivate governments, local people, and international organizations to conserve biological diversity. Incentives are used to divert resources such as land, capital, and labor towards conserving biological resources, and to facilitate the participation of certain groups or agents in work which will benefit these resources. A perverse incentive is one which induces behavior which depletes biological diversity, though of course such perversity is in the eyes of the beholder. A disincentive is any inducement or mechanism designed to discourage governments, local people, corporations, and international organizations from depleting biological diversity.

Economic incentives can take a large number of forms and can be categorized in several different ways. A taxonomy of the various sorts of incentives is presented in Box 6, along with examples of the sorts of incentives that might be relevant at community, national, and international levels. This Table could have been much more complex and detailed, but it is sufficient to indicate the major headings under which the various types of economic incentives can be placed. Some of the categories would fit as easily in one place in the taxonomy as in another, and the most effective incentives frequently fall in more than one category. Most incentive packages would contain a mix of these.

Disincentives include taxes, fines and penalties of other types (which are usually administered through legislation) as well as public opinion or peer pressure (the use of which is far more subtle). Together, **incentives and disincentives provide the carrot and the stick for motivating behavior that will conserve biological resources.**

A major objective of using incentives is to smooth the uneven distribution of the costs and benefits of conserving biological resources; rather than suppressing the symptoms of resource misallocation, they are intended to address the cause of such abuses through providing a means of reaching compromise on substantive environmental conflicts (Sorensen, *et al.*, 1984). They preserve the status quo by mitigating anticipated negative impacts on local people by regulations controlling exploitation of biological resources, and compensate people for any extraordinary losses suffered through such controls. They also improve the status quo by rewarding the local people who assume externalities through which the larger public benefits. Finally, they can open up the decision-making process to the people who are most directly affected by conservation of biological resources.

In correcting market failures, incentives provide a policy tool for overcoming the major constraints to conservation activities such as reforestation, and wildlife and protected areas management. They can convince villagers, industry, and government of the benefits of such efforts, provide the financial means to implement them, the legal support for addressing land-tenure or land-use problems, and the financial and technical capability to develop productive systems which do not deplete biological resources.

Box 6. Examples of Economic Incentives for Conserving Biological Resources

Type of Incentive	*Examples*		
	Community	*National*	*International*
I. *DIRECT INCENTIVES*			
1. In Cash	Subsidies for reforestation	Research grants	World Heritage Fund
2. In Kind	Food for work in a reserve	Forest concessions	WWF equip. for pandas
II. *INDIRECT INCENTIVES*			
1. Fiscal Measures	Compensation for damage by wild animals	Price support for intensive agriculture	Commodities agreements; debt swaps
2. Provision of Services	Community development	Conservation education	Technical assistance
3. Social Factors	Enhanced land tenure	Training for staff	Intl. data bases

---

Incentives are clearly worthwhile if they can stimulate activities which conserve biological resources, at a lower economic cost than that of the economic benefits received. But proposals for incentive packages must also demonstrate cost/benefit ratios higher than the ratios for other proposals competing for scarce capital; this is another reason to demonstrate as thoroughly as possible the economic value of conserving biological resources.

The approach taken here is to begin the analysis at the community level (Chapter 4), because this is where most incentives must have their impact. Incentives must serve to correct problems perceived by people in the vicinity of areas which are of particular importance for conserving biological resources, developed at the community level, and applied within the context of local social organizations. However, for incentives to function well at the local level, they need to be supported by appropriate policies at the national level (Chapter 5). Additional incentives are also relevant at the national level, for both the government institutions involved in managing biological resources and the institutions whose activities frequently involve external effects on biological resources. Finally, economic incentives often require support from the international community (Chapter 6), primarily in terms of technical assistance, information, commodities agreements, and various measures reflecting existence value as perceived by the international community; these are most relevant for developing countries, but Case Study 23 shows that they can also be important in developed countries. The question of how to pay for incentives packages is addressed in Chapter 7.

FORM AND FUNCTION OF THE DIFFERENT TYPES OF INCENTIVES

Direct Incentives

Direct incentives are applied to achieve specific objectives (e.g., to reduce poaching of protected wildlife, to improve management of a protected area, to promote sustainable utilization of forest resources). Direct incentives can be either in cash or in kind, but in any case should be conditional on changed behavior toward biological resources. Direct incentives are often linked to specific rewards; direct income supports to farmers, for example, can be linked to a program of land retirement on environmentally sensitive lands.

Direct Incentives in Cash. Not surprisingly, direct cash subsidies are often the most welcome, since they can be used in the most flexible way. They include fees, royalties, rewards, grants, income supports, subsidies, loans, and daily wages. Such cash awards function as incentives only when they are clearly and overtly linked to changes in behavior, and specifically toward behavior which conserves biological resources. Cash disincentives include penalties and fines. The major problem with direct cash incentives is that they may produce long-term disincentives to conserving biological diversity by creating a dependency on outside aid. The proper use of cash incentives is to provide those affected with a sense of empowerment and responsibility for their own destiny; cash incentives which promote self-sufficiency with minimal dependence on outside aid and inputs should be favored.

Direct Incentives in Kind. Direct incentives in kind include material goods which are delivered to institutions, communities or individuals in return for their contribution to biological resource conservation and rehabilitation works; or in return for their refraining from activities which damage biological resources. Other direct incentives in kind include food-for-work programs, equipment donated to protected area management authorities, timber concessions (accompanied by appropriate conditions on extraction), and providing access to certain protected resources under certain conditions (often in buffer zones around protected areas). Direct in-kind disincentives might include jail sentences, confiscation of land or elimination of use-rights, as in mandatory retirement of marginal land.

Indirect Incentives

Indirect incentives encourage behavior which conserves biological resources or generate resources for conservation efforts without any direct budgetary appropriation for biological resource conservation from the government or other sources. They involve applying fiscal, service, social, and natural resources policies to specific conservation problems and may involve providing preferential treatment in trade agreements, price supports, or land tenure.

Fiscal measures. Fiscal policy is concerned with gathering income to meet public expenditure in support of conserving biological resources, complementing economic policy measures which promote investment, production, and employment related to sustainable use of biological resources. Fiscal incentives are a legal and statutory means of channelling funds towards conservation activities, involving such indirect measures as tax exemptions or allowances, insurance, guarantees, tariffs, and price supports. At the international level, so-called "debt swaps" and foreign assistance projects can provide fiscal incentives to governments.

Provision of services. When a government has decided that certain biological resources or areas are of outstanding value to the nation as a whole, it should consider what sorts of services it might be able to provide to the communities most directly affected by any restraint on use. As incentives for changing their behavior regarding the biological resources to be protected, such communities can be provided with accelerated development activities in recognition of their contribution to national objectives in conserving biological diversity. Governments may decide that public opinion could be so important in promoting conservation of biological resources that it is willing to invest in major public education programs, and international agencies may decide that investing in technical assistance for biological diversity projects is worthwhile.

Social factors. Social incentives are designed to improve the quality of life of the community or nation, ensuring that benefits from biological resources are equitably distributed. They include a wide range of measures aimed at developing a harmonious and sustainable relationship between people and biological resources, including enhanced land tenure, training and education, employment in activities related to biological resources, building up of institutions to manage biologial resources, and information on the status and trends of biological resources.

COSTS AND BENEFITS OF INCENTIVE SYSTEMS

While systems of incentives and disincentives can efficiently promote the conservation of biological resources, some costs are involved in their administration. All incentives require some degree of regulation, enforcement, monitoring, and feedback if they are to function effectively. Subsidies to game ranches or sea turtle hatcheries need to be administered, voluntary labor needs to be supervised, collection of stumpage fees requires field staff, and enhancing land tenure may require major investments in cadastral surveys, legal fees, and registration.

Further, economic incentives and disincentives must be used with considerable sensitivity if they are to attain their objectives, and they must be amenable to modification and adaptation to changing conditions. Gate fees to national parks, for example, may need to be increased to keep up with inflation, and gate fees to national parks which receive few visitors may need to be discontinued if they yield less revenue than it costs to collect the fees.

The concept of Marginal Opportunity Cost is extremely helpful in determining when economic incentives are being used appropriately. For example, a subsidy may be said to exist if the price paid for publicly provided goods or services is less than MOC. In this regard, Warford (1987a) has pointed out that "it will often be the case that increasing prices beyond those required to meet the financial objectives of power utilities will improve the efficiency of resource utilization and do so in a way that enhances environmental objectives."

The kinds of empirical and conceptual problems encountered in designing a system of economic incentives to conserve biological resources are very similar to those related to the conduct of benefit-cost studies (Warford, 1987b). The cost of implementing the incentive system should be compared with the estimated benefits in terms of conservation of biological resources. The magnitude of the savings would depend upon the reaction of the users to the incentives and disincentives and the MOC of the activity to which the incentive scheme is formally applied.

THE PROBLEM OF PERVERSE INCENTIVES

Introduction

The preceding discussion has shown that economic incentives can play a very major role in promoting more effective conservation of biological resources; chapters 4, 5, and 6 and the case studies will provide further evidence that this is so. However, to date economic incentives have been far more pervasive in over-exploiting biological resources than conserving them. An economist at the World Bank has identified one of the main problems: "In developing countries, the relevant decisions are frequently made by a small, politically influential group with interests in commercial logging, ranching, plantation cropping, and large-scale irrigated farming operations. As a result, the prevailing systems of investment incentives, tax provisions, credit and land concessions, and agricultural pricing policies tend to favor those in power, causing losses for the economy as a whole, and at the same time damaging the environmental and natural resource base" (Warford, 1987b).

Any subsidy has the effect of lowering market price and thus making the gap between social and market price even wider; a subsidy on a resource will cause more of the resource to be demanded, because the price is lower than it otherwise would be. If the use of that resource already generates external costs for the environment, then the subsidy will make things worse.

Sound economic planning would involve maximizing total benefit from *all* possible direct and indirect uses of biological resources over the long run, accommodating the needs and values of all interest groups, whether or not those values are reflected in market transactions. "Government policies frequently violate this criterion," says Repetto (1987a). "By manipulating tax codes, public credits, and charges for the use of public lands, they typically create fiscal burdens for taxpayers, while sacrificing long-term economic welfare and wasting forest resources." Improving such policies can enhance long-term economic benefits, provide more effective conservation of biological resources, and reduce fiscal burdens on governments.

Major Economic Incentives for Depleting Biological Resources

As suggested in Chapter 2, the world is replete with examples where unsustainable uses of biological resources have been justified by arguments based on economics. In most parts of the tropics, the opening up of forest areas is supported by powerful economic incentives in the form of state-sponsored road-building programs which facilitate access to markets and thereby increase potential profits from converting forest to agriculture or grazing. Further, resettlement of poor people in the remote forested areas opened up by roads is often politically preferable to genuine land reform which involves the redistribution of existing agricultural lands.

Many incentives aimed at stimulating production have significant external costs. For example, in China, the harvest of musk from wild musk deer is stimulated by high prices offered by the Department of Primary Production; but the snares set for musk deer also capture giant pandas, snow leopards, and other protected species (and indeed the musk deer itself is a protected species). When government policies conflict, the cash incentive often outweighs the disincentive of fines or jail sentences—in the Chinese case, killing a panda can carry the ultimate disincentive of a death sentence for the poacher, but pandas are still being killed in snares set for musk deer.

Examples of perverse incentives could be drawn from virtually anywhere, but the following suggest the kinds of problems that have arisen.

Schumann and Partridge (1986) have presented numerous case studies demonstrating that Latin American governments and international development agencies have tended strongly to support policies which encourage land settlement in tropical forest areas, through road construction and other forms of subsidy. Converting coffee estates to cattle ranches has increased unemployment in the highlands of Chiapas, Mexico, thereby encouraging many peasants to settle and clear new land in the forested lowlands (in turn depleting biological resources); and mechanised soybean production in Brazil and Paraguay has displaced many small farmers, who have moved on to settle in previously forested areas. Ledec and Goodland conclude that governments wishing to settle their forested frontiers may consider it desirable to *reduce* employment options on existing agricultural lands, thereby providing a perverse incentive through ensuring an ample pool of settlers willing to risk the hardships of frontier life in order to make a living. Further, in many parts of Latin America, landowners or land claimants who do not clear the forest often risk losing title or other legal rights to the land (see Case Study 1).

The implications of such policies for biological resources are apparent. But the package of incentives for forest conversion is justified by governments which suffer from over-crowded cities and are blessed with a sparsely-populated hinterland, as a sacrifice which will generate capital to support development in the newly-settled lands.

The Government of Indonesia has a similar problem, being faced with severe over-crowding in Java, Madura, and Bali (Java alone has 100 million people living on a land area equivalent to New York State or Greece). Its transmigration program seeks, as a policy objective, to move poor farmers from these inner islands to settle areas in the outer provinces of Sumatra, Kalimantan, Sulawesi, and Irian

Jaya, which are currently under forest and occupied by sparse populations of shifting cultivators. Such policies incorporate various economic incentives to clear forest land, thereby reducing biological diversity. In effect, rising population has forced the Indonesian Government to convert its wild forest capital into uses that it hopes will provide durable benefits to larger numbers of people.

Using subsidies to intensify agriculture in Indonesia's more densely-population areas has also caused negative impacts on the environment. For example, subsidies on pesticides have led to their over-application, with consequent pesticide poisoning incidents (one causing 18 deaths from a single village), loss of insect predators (which means reappearance of the pests), toxic effects of fisheries, and the breeding of "super-pests." When a number of pesticides were banned in 1986, it was quickly discovered that alternative, integrated approaches to pest management were far more effective anyway; the government is now examining incentives for promoting integrated pest management instead of over-use of pesticides.

In Botswana, the government provides agricultural subsidies for the full costs of plowing (up to 10 hectares), together with additional subsidies for row planting and weeding; for "destumping," clearing land for cultivation; for the full cost of seeds; and for fencing. It is clear that the full-cost plowing subsidy along with the free distribution of seed provides a very strong incentive for mixed farmers to plant an area in food crops well in excess of the expected harvested area; the ratio of harvested to planted area therefore averages less than 50 percent. The destumping subsidies also contribute to the devegetation of arable lands, but the fencing package may be more significant for grazing lands. The promotion of wire over traditional thorn fencing qualitatively and quantitatively changes the timber demands of fencing; wire fencing requires posts that can only be obtained from larger trees, while the effect of termites means that posts cut from most species require regular replacement; the result is that quality of rangeland declines through elimination of tree cover and the encouragement of bush encroachment (Perrings, *et al.*, 1988) (see Case Study 2 for additional examples).

These examples incorporate economic justifications for reducing the stock of biological resources; reducing the flow of environmental services is often an external effect of such policies. This essentially permanent drawing down of natural capital is justifiable in economic terms if it provides sustainable benefits which exceed the benefits of conserving these resources. The problem is that such incentive systems have too often led to permanent degradation of resources rather than their enhancement, causing significant long-term economic losses to governments. Incentive schemes to boost agricultural production, for example, can contribute to problems of soil erosion, deforestation, and water scarcity. By promoting the extension of agricultural land, they can thereby deplete biological resources in natural habitats and by spreading dominant market crops, they can reduce the diversity of cultivars and so-called "minor crops."

The accompanying case studies on incentives for land clearance in Amazonia (Case Study 1) and forestry in several tropical countries (Case Study 3) demonstrate that incentives for forest clearing have too often led to permanent depletion rather than sustainable development.

Commodities Trade and Incentives to Over-exploit

Virtually all civilizations have been founded on trade. In today's world, no nation can maintain or improve its standard of living unless it is able to call on raw materials and manufactured goods from distant lands. This international trade has enabled the entire world to draw on the globe's collection of local ecosystems, feeding the growing population of the world at the expense of sometimes over-exploiting local resources. International trade requires foreign exchange, which most tropical countries earn from the exploitation of natural resources; this has inevitably led to increased consumption of biological resources, sometimes at rates that cannot be sustained.

The tropical nations are if anything more economically dependent on trade with industrialized nations than vice versa, and the lack of stability in the markets of the primary commodities from these countries makes it difficult for them to ensure the sustainable utilization of their biological resources. Yet the biological diversity they contain is a global asset; to conserve this asset at the global level requires significant change in the trade and international relations policies of industrialized nations. Progress is already in evidence: aid agencies in industrialized countries are beginning to channel support into sustainable development activities, conservation of biological resources, and so on. But the potential of this support remains limited by the trade and foreign policies of these same industrialized countries.

International trade in agricultural commodities has major impacts on national efforts to use biological resources sustainably. For example, Botswana exports more than half its beef output overseas. Two-thirds of it goes to the EEC, which has a high demand for lean (grass-fed) beef. So EEC development aid programs heavily subsidize the beef-export business in Botswana, where cattle are rapidly replacing wildlife. Wildebeest have dwindled in numbers through loss of habitat until they now total only 10 percent of their once vast number. A similar process of wildlife depletion, with the consequent degradation of rangelands by overgrazing of domestic cattle, can be seen in several other African countries where the spread of ranching is subsidized by foreign aid.

Despite the cyclical drought, Sahelian countries continue to grow more agricultural produce each year. But most of it is commodities for export, not food for local consumption. In 1984, Sahelian countries harvested almost seven times as much cotton as in 1961, and they imported almost nine times as much cereal grain. The cash-crop trade is supported by virtually all Sahelian governments on the grounds that it earns foreign exchange to buy manufactured goods from the industrial countries, attracts support from development aid programs, and brings commercial investment from Europe. The expansion of cotton and peanut production has driven many subsistence farmers onto marginal lands too dry for farming, turning semi-arid land to desert with often disastrous results for both humans and the biological resources that support them.

Government policy imperatives to earn foreign exchange have often led to incentives which have had the effect of over-exploiting biological resources and reducing the economic viability of sustainable forms of development. In Brazil, for example, the Superintendency for Fisheries Development provided more than $100 million from 1967 to 1973 to private firms at very low interest rates, but only to firms servicing the export market. Artisanal fisheries, which contributed more than 50 percent of the fisheries catch, did not receive any incentives. As a result, lobster, shrimp, and catfish were over-harvested, and the industrial fleet from the rich southern states, after destroying the stocks in the area, moved to the productive areas of the Amazon basin, repeating the same overfishing. These incentives resulted in the depletion of biological resources and led to the impoverishment of artisanal fishermen (Diegues, 1987).

Investments in conservation projects by aid agencies and NGOs are welcomed by governments throughout Africa, but most of them are doomed to be overwhelmed by the more powerful trade imperatives.

All this is not to say that exports of commodities from the developing world to industrial countries should be discouraged. In fact, developing countries have too often penalized commodity exports through disincentives such as overvalued exchange rates, port taxes, and levies imposed by crop marketing boards, thereby severely restricting returns to farmers as well as stifling production urgently required to earn foreign exchange. The main problem with export crops is that they are no longer supporting the local human needs, but rather feed the demands of distant markets (and often require importing energy in the form of agricultural chemicals). Because they are not responding to local conditions of supply and demand, no real limit is placed on the impact that such crops can have on delicately-balanced agricultural systems which have developed over long periods of time.

On the other hand, nothing about exportable crops makes them inherently more damaging than subsistence crops; many export crops are perennials and tree crops, which, when grown with grasses underneath, afford better protection against soil erosion than row and root crops such as cassava, maize, and millet. But export crops are part of the international economic system rather than the national or local economic system, so costs and benefits are shared in ways that are quite different from subsistence crops. Provided that appropriate incentives are designed and implemented for ensuring that the traditional staple crops are also produced, that export earnings are allocated equitably among the rural people, and that agricultural development is balanced with efforts to conserve biological resources, export crops can be important factors in the overall development of many countries.

Trade in commodities, including biological resources, is used to build the capital necessary to invest in modernizing the economy, an approach all the more justifiable in developing countries with young economies. Yet protectionism in the richer countries sharply reduces the ability of developing countries to generate adequate income from the use of their natural resources, causing needless depletion of biological resources and a reduction of the capacity of these resources to fuel future development. Moreover, the prices of commodity exports do not reflect the environmental costs of sustainable management and use of these resources. In a sense, the developing countries subsidize importers of their products, incurring important short-term and especially longer-term costs to themselves and their environments, and compromising their development prospects. Add to this the unpredictable nature of trade regulations, where new restrictions can often undermine investments which were sensible in the context in which they were made, and it becomes clear that developing countries must navigate a treacherous course in seeking to build sustainable economies through international trade in biological resources.

Sustainable development which includes effective management of biological resources will require a vigorous effort to negotiate and implement international commodity agreements which would provide economic incentives to stabilize and maximize the earnings of developing countries from exports of primary products, thereby enabling them to evolve a more balanced development base and manage their biological resources on a sustainable basis.

However, to date commodity agreements have been no more than another form of protectionism, and to the extent that they artificially drive up prices for the commodities involved, are more likely to stimulate over-exploitation than conservation.

As a result, efforts at the individual project or country program level which aim at providing incentives to conserve biological resources must take account of the trade and aid imperatives that might be working against the conservation effort. This argues for integrating the "biological diversity projects" with other efforts, rather than attempting to implement them as symbolic efforts at conservation.

Modifying Perverse Incentives

In the sense used here, incentives are perverse when they stimulate behavior which tends to deplete biological resources. Governments have often instituted these perverse incentives for important political or social reasons, and the impact on the environment is often an externality. Agricultural incentives, for example, are exceedingly difficult to reduce once they have been established, irrespective of how perverse they might be for biological resources.

In such cases, it may be necessary for governments to institute new conservation incentives which effectively cancel out the negative impacts of perverse incentives; in effect, governments are paying twice for something for which they would not have had to pay at all if their policies were environmentally sound in the first place.

Governments finding themselves in such an uncomfortable situation should consider the extent to which the widespread use of subsidies has led to increasingly negative sectoral and cross-sectoral impacts, especially in agriculture. Heavy subsidies are becoming a major constraint not only to the viability of the agricultural sector itself, but also to the responsiveness of the development budget as a whole, especially in a period of static or declining government revenues. On the other hand, price controls on agricultural commodities often serve as disincentives to conservation of cultivated land; easing price controls can serve as an incentive to invest scarce resources in research and development and to adopt new technologies.

While the details will vary from place to place, the dual problems of subsidies and price controls on agricultural commodities occur widely throughout the tropics. The best solution would appear to be diversified farming systems which are in tune with local ecological conditions, and which are based on locally-available resources to the maximum extent possible. Tarrant et al. (1987) report that in Indonesia, input subsidies, particularly for fertilizer, pesticides and irrigation are imposing considerable external costs in terms of agricultural pollution and resource depletion. They question whether a production-led approach is suitable for the diversity of agro-ecological systems that characterize Indonesia's marginal lands, pointing out that failure to consider farming and cropping systems as the basis for agricultural development strategies means that many traditional agroforestry and home garden systems are not being adequately developed.

More integrated agro-ecosystems or farming systems approaches would require a greater investment in research, marketing infrastructure and extension; nevertheless, this could be at least partly financed by a reallocation of funds from the removal of pesticide subsidies, a gradual removal of fertilizer subsidies, an effective system of water charges (e.g., increased taxes on irrigated lands) and the removal of credit subsidies to sugarcane.

Instead of the subsidies on cattle ranching in Amazonia, a tax levied on livestock production might reduce over-grazing, and lead to a reduction in land clearance. By reducing the rate of soil erosion, imposed forest protection would exert a beneficial influence on agricultural productivity many kilometers away. Ideally, the tax should be such that the livestock producer faces total input costs equal to the the MOC of his activity, which is determined by such factors as the effects on soil erosion and consequent impact on agricultural output elsewhere in the system.

In Botswana, altering beef prices could improve land use, by changing the seasonal margins to encourage more offtake in communal areas; reducing intergrade margins to encourage more offtake of lower quality grades; and raising agents' margins to encourage greater offtake. Other steps which could help correct the current system of perverse incentives could include levying a "management fee" or range rental that varies inversely with rainfall, to reflect the inverse relation between user costs and rainfall; introducing water charges which reflect the scarcity of the resource; modifying the tax benefits available on livestock; establishing producer prices at levels that encourage an increase in offtake; and subsidizing voluntary reduction in herds in areas where range degradation has already reached severe proportions (Perrings, *et al.*, 1988).

NEGATIVE IMPACTS OF ECONOMIC INCENTIVES PROVIDED BY THE INTERNATIONAL COMMUNITY, AND WHAT CAN BE DONE TO OVERCOME THEM

Perverse incentives are not the only ones with negative impacts. Even effective and well-meaning incentives can have unintended ripple effects which can bring results contrary to those intended. An impressive range of development agencies, government institutions, universities, and NGOs have provided economic incentives to countries in the tropics to conserve biological diversity. But most of these incentives have also led to increased burdens for the relevant government institutions, and many of them may have caused as much harm as good. "Most support from the international community for conservation projects in developing countries concentrates on research, planning, and development functions of conservation agencies," Barborak (1988a) concluded on the basis of over a decade of experience in Central America. "Often such projects greatly increase the need for long-term operational funding and personnel levels by the recipient agency in an era of generally declining real manpower and budget levels. . . . Most sources and forms of outside support do not adequately address the need for greater self-sufficiency, stability in personnel and operating budget levels, and a secure long-term financial footing based primarily on local resources."

In Sri Lanka, for example, a USAID-supported project to develop a system of protected areas as part of a water resources development scheme in the Mahaweli region drained resources from the other protected areas in the system, and will create an infrastructure that is far beyond the capacity of the Department of Wild Life Conservation to maintain under its current budget. Many, even most, externally supported projects which have prepared management plans have been little more than displacement behavior, with the plans forgotten as soon as the expatriate advisor has departed. The millions of dollars which have been spent on field research projects in developing countries by scientists from developed countries have yielded relatively little information that is actually used by the responsible government agency.

What, then, can be done to minimize the negative impacts of economic incentives from international sources, and turn them into positive incentives which have lasting value? The basic principles should be the following:

- Develop local capacity to manage biological resources; strive for self-reliance rather than dependence, including helping to design sustainable sources of income for supporting personnel, equipment, and maintenance.
- Contribute to ensuring the long-term economic visibility of protected areas, including designing systems of sustainable utilization of biological resources.
- Contribute to projects which are designed specifically to meet the needs of the management agency, not to attempt to mold that agency into a temperate-zone model.
- Enable conservation agencies to participate fully in regional and global conservation efforts, including the various conventions and treaties relevant to biological resource management.
- Treat both the symptoms and the causes of depletion of biological resources, recalling that the causes may be very far removed from the resource management agency.

- Promote the integration of biological resource conservation into the larger development issues of the country, based on the principles of sustainable development.

CONCLUSIONS

Major improvements in conserving biological diversity can be made at the policy level by various government agencies through the use of economic incentives and disincentives. Such incentives— at community, national, and international levels—need to be included as part of a package of government policies which address issues of rural development, research, education, training, resource management, legislation, and institutional development.

Economic incentives have been used by governments to open up their frontiers to settlement or otherwise stimulate high production from biological resources. This has resulted in conversion of forests and other wilderness to a range of agricultural uses and the depletion of biological resources. While using such incentives may have been appropriate when biological resources were plentiful, the process is reaching its productive limits (and indeed has exceeded them in many places). A major step in moving from exploitation to sustainable use is for governments to analyze the impacts of all relevant policies on the status and trends of biological resources. Such an analysis would involve detailed determination of Marginal Opportunity Costs, including costs and benefits of direct and indirect values.

Based on the policy review, governments should eliminate or at least reduce policy distortions such as subsidies that favor environmentally unsound practices, and at the same time discriminate against the rural poor, reduce economic efficiency, and waste budgetary resources. Overcoming the damage caused by perverse incentives will require new incentives to promote conservation, applied at a series of levels and in a number of sectors.

Any incentives need to be designed with great care and applied in ways that will ensure that they will attain their objectives. Poorly designed incentives can easily backfire. Long-term loans, for example, may be used as incentives to deplete biological resources as well as to conserve them; a subsidy on selective logging may well discourage clear-cutting, but it may also encourage forestry activities over a larger area and thereby negate any benefits that may have been gained for biological diversity. Therefore, incentives must be designed specifically to achieve the objectives for which they are intended, and measured by that criterion.

Finally, incentives need to be protected from over-success. It may sometimes happen that the incentives package for an area will be so attractive that it draws in rural people from other areas, thereby possibly negating any benefits that are gained from the incentives. Incentives therefore need to be finely tuned to the Marginal Opportunity Cost relevant to the communities involved.

CHAPTER FOUR

THE USE OF ECONOMIC INCENTIVES TO PROMOTE CONSERVATION OF BIOLOGICAL RESOURCES AT THE COMMUNITY LEVEL

INTRODUCTION

Biological resources by definition have real or potential benefits for humanity. The specific package of available biological resources varies considerably from place to place, depending on such factors as evolutionary history, soil, rainfall, and history of human use. The people who live closest to the resources have often developed specific ways and means of managing these resources. For the people living in or near the forests, plants and animals provide food, medicine, hides, building materials, income, and the source of inspiration; rivers provide transportation, fish, water, and soils; and coral reefs and coastal mangroves provide a permanent source of sustenance and building materials.

Biological resources are often under threat because the responsibility for managing resources has been removed from the people who live closest to them, and instead has been transferred to government agencies located in distant capitals. But the costs of conservation still typically fall on the relatively few rural people who otherwise might have benefited most directly from exploiting these resources. Worse, the rural people who live closest to the areas with greatest biological diversity are often among the most economically disadvantaged—the poorest of the poor.

This is both unjust and a recipe for undermining any conservation efforts the government may design, and that development assistance agencies may support. Under such conditions, the villager is often forced to become a poacher, or to clear national park land to eke out a crop. Changing his behavior requires first examining government resource management policies, land tenure, agricultural prices, and many other policies which may stimulate a villager's poaching and encroachment. Economic incentives designed for reversing these policies at the community level may provide the best means of transforming an exploiter into a conservationist.

Measures designed to induce prudent management of biological resources by local people may include assigning at least some management responsibility to local institutions, strengthening community-based resource management systems, designing pricing policies and tax benefits which will promote conservation of biological resources, and introducing a variety of property rights and land tenure arrangements. These incentives and disincentives may serve to rekindle traditional ways and means of managing biological resources which have been weakened in recent years due to economic pressures at the national and international level. Providing local communities with viable alternatives for earning a living will also require educational opportunity, equitable land tenure, and access to credit so that decision-makers at the household and small farm level are able to respond effectively to incentive systems.

THE USE OF ECONOMIC INCENTIVES AT THE COMMUNITY LEVEL

Changing behavior at the local community level may not be as difficult as it sounds. Poaching and illegal shifting cultivation are hard work, uncomfortable, risky or even dangerous, and are often only marginally profitable, so many villagers will willingly adopt more sustainable ways of earning a living if they are given the opportunity to do so (MacKinnon *et al.*, 1986). Most villagers realize that **they are better off when they manage their biological resources in a sustainable manner than if they deplete them**. They may sometimes find it necessary to exceed sustainable yields for short periods of time, if the harvest is converted into durable capital which in turn supports sustainable development; trees cleared from agricultural land, for example, can be sold as timber which earns sufficient profits to enable the farmer to build terraces on his agricultural land. But in general, villagers want to conserve, and seeking greater conservation action from a villager requires that real benefits be provided to him or her, often in the form of alternative sources of income.

Since the actual management of biological resources is inherently dispersed and decentralized, the traditional systems of resource management used by local people were often far more effective and sustainable than the new systems devised by governments. Central government control has seldom been accompanied by sufficient resources—including funding, trained personnel, and political will—to ensure that the remaining biological diversity is indeed conserved as effectively as it was before development projects penetrated the remote areas. Even where sufficient resources were once available, the economic conditions of the late 20th Century dictate that alternative approaches be applied to ensure the survival of the biological diversity on which the prosperity of human communities depends.

When the central government assumes stewardship for biological resources of national importance (in effect establishes a monopoly over areas of outstanding importance), then it needs to use economic incentives to encourage the local people to respect the new regulations. Such incentives can include grants, accelerated development aid, education, improved health care, and a whole range of other mechanisms to compensate villagers for any losses they may suffer from being denied resources that were previously "theirs."

Such incentives packages are likely to function best when they are implemented as part of larger rural development efforts, requiring considerable cooperation between government agencies involved in conservation, government and private agencies responsible for development, and local community-based organizations.

Incentives to conserve biological resources at the village level can address a number of objectives:

- to build the capacity of communities adjacent to protected areas to develop productive activities which do not deplete biological resources.
- to reduce agricultural pressure on marginal lands, which are often better devoted to conserving biological resources, whose

abundance and/or quality has been increased as a result of protection.

- to concentrate agricultural development on the most productive agricultural lands (those best able to respond well to yield-increasing technology).
- to conserve traditional knowledge about the use of biological resources, and the cultural systems which hold such knowledge; and to re-establish common property management institutions where these have been effective in the past.
- to compensate villagers for possible income lost through restrictions on utilization of protected biological resources; or for damages suffered from the depredations of wild animals on crops or livestock.

Most of the case studies deal with incentives at the community level; a summary is presented in Box 7.

DIRECT INCENTIVES TO PROMOTE CONSERVATION OF BIOLOGICAL RESOURCES AT THE COMMUNITY LEVEL

In order to compensate villagers for lost resources, they may be provided directly with cash in various forms or given access to some of the biological resources of the protected area, including such things as building materials, thatch grass, meat and other animal products. A variant of this approach is to provide local communities with the profits from both consumptive and non-consumptive uses of the protected area. The use of such direct incentives will vary with the situation, and must be designed to meet the specific objectives that are defined.

Direct Incentives in Cash

In times when cash is chronically short in most government budgets, creativity is often required to generate the necessary money to provide direct incentives to local communities, since the regular budgetary resources of governments are seldom adequate to do a fully effective job. Cash incentives can often be generated by the controlled harvesting of protected biological resources on government-owned or communal land. For example, in Zimbabwe, "Operation Windfall" provided the proceeds of elephant culling—including meat, skin, and ivory—to two local councils to use for community

Box 7: Community Incentives and Disincentives Addressed By the Case Studies

Case Study	Disincentives	Community organization	Community development	Access to resource	Land tenure	Training	Education	Employment	Fiscal
4. Thailand's Khao Yai National Park	X	X	X		X	X	X	X	X
5. Thatching Grass, Zimbabwe		X		X					X
6. Brazil's Iguape Estuary	X	X		X					
7. Sagarmatha National Park, Nepal		X	X	X	X		X		
8. Amboseli Ecosystem, Kenya			X	X				X	X
9. Marine Resources, Quintana Roo, Mexico	X	X		X					
10. Food for Work, Wolong, China			X			X	X	X	
11. Hunting & Local Communities in Zimbabwe	X	X	X	X					X
12. Inner Delta of the Niger, Mali		X	X	X					
13. Tegucigalpa Watershed, Honduras	X		X			X			
14. Coastal Zone of Japan	X	X		X					
15. Pastoralism in Northern Kenya	X	X	X		X		X		
16. Marine Resource Management, Philippines	X	X		X		X			
18. The Rubber Tappers' Movement in Brazil	X	X			X	X		X	
19. Villages & Wildlife Reserves, India			X						
20. Revolving Fund in Zambia	X	X				X		X	X
24. World Heritage Subsidies in Australia			X						X

(Note: Only case studies addressed at the community level are considered.)

development projects. Between early 1981 and June 1982, the councils received $960,000. Elephant poaching dropped so dramatically that the parks department found it no longer necessary to post wardens in the area (Martin, 1986). Cash awards are also the most difficult to control, once they have been awarded; it is therefore often better to provide cash to reward past actions rather than in anticipation of future action.

Fees. Since areas of national importance for conserving biological resources are usually of outstanding interest to visitors, it is often reasonable to charge fees for entering protected areas. Portions of these fees can be returned to nearby villages as an incentive for maintaining the integrity of the reserve.

Rewards. Cash rewards can be provided to villagers for outstanding service, or for exemplary behavior regarding conservation of biological resources (the obvious difficulty is judging such behavior). Rewards can also be provided for informing the authorities of illegal harvesting of biological resources (in other words, helping to implement disincentives). Such actions need to be carried out with extreme sensitivity, and only when a serious problem exists and where the larger community is in support of the effort. Reward schemes are often most effective in seeking support from the local community for controlling the activities of outsiders.

Fines. To serve as an effective disincentive, fines on offenses such as illegal cutting of trees or poaching of animals need to be higher than the value of the tree or animal poached. Fines can also serve as incentives, if a portion of fines is returned to local villagers for development activities and village leaders are assigned policing responsibility; this would clearly work better among some ethnic groups than among others. A major drawback of fines is that enforcement is imperfect, and poachers consider both the size of the penalty and the probability of getting caught. To discourage poaching, an extremely high fine may have to be imposed (such as the death penalty recently enacted by China for poaching giant pandas), or enforcement increased; very high fines are unlikely to be taken seriously by rural people who are so poor that they need to poach.

Compensation. Where protected animals prey on livestock or feed on field crops, or kill or injure villagers (a common occurrence in many countries with large predators like lions, tigers, or crocodiles, or dangerous herbivores like elephants, rhinos, or buffalo), cash compensation for damages may be a necessary disincentive, if large-scale punitive raids on the offending wildlife are to be avoided. In Sri Lanka, for example, farmers who suffer from elephant damage to their crops can be paid for contributing information and other support to elephant capture operations, with their compensation coming from the profits from auctioning the new captives.

Grants are normally provided on the basis of proposals for specific activities, usually of a relatively short duration. Grants are most effective when they generate a necessary change in behavior, provide the foundation for a change which is then sustainable, or to build the capacity to benefit from other types of incentives. The poorest villagers will seldom have access to grants because the application process is likely to be beyond their capacity; but national NGOs can often assist in this process.

Subsidies involve the outright provision of financial assistance (a capital grant), usually by the State; in essence, they are negative taxes to support those activities which intentionally or necessarily operate at a loss—possibly due to market failure—while still meeting community needs. Subsidies are usually used where financing at commercial rates is not available for activities of relatively long duration that require significant levels of investment, such as reforestation, development of plantations, and development of butterfly ranches, crocodile farms or wildlife management schemes (which contribute to conservation when they reduce pressure on wild stocks, or when they provide incentives for maintaining wild stocks as sources of breeding material). Subsidies may also be granted for avoiding activities which damage biological resources, such as keeping cattle out of a national park or preventing goats from damaging young reforestation projects. They may cover all or part of the cost of such activities, but typically are aimed at generating investment from other sources as well.

Land banks. When an objective is to reduce the amount of land under agriculture, and thereby increase the amount of land for which conservation of biological resources is a primary objective, direct income supports can be linked to a program of land retirement on environmentally sensitive lands. This is particularly appealing to farmers (and especially shifting cultivators) who are forced by income insecurity to cultivate marginal lands.

Loans (or credit) are often useful for supporting activities aimed at conserving biological resources and which require funds beyond the capacity of an individual to provide. Loans may be sought from commercial banks or other sources (for some activities, at least part of the interest may be subsidized). Loans may reduce pressure on protected biological resources by enabling the community to prosper on its own land, through providing better access to markets, improving packaging, and negotiating better prices for products. An important supporting measure is for banks to consider plantation trees as collateral, in the realization that trees will mature into productive assets.

Revolving funds are a type of loan where an initial capital fund is established in a community to provide short-term loans to villagers to purchase inputs and to hire labor for short cycle crops. The loan is repaid, with nominal interest, after harvest, thereby providing the basis for another cycle of loans. Revolving funds are normally operated by a community organization, which gains in prestige and influence as the fund provides villagers with necessary assistance. The manager of a revolving fund must be well-motivated, well-trained, and well-respected by the community.

Daily wages are paid by the conservation agency to individuals or community organizations in return for activities which contribute to conservation of biological resources (reforestation, soil conservation, farming of traditional crop varieties, construction of trails or firebreaks within a national park, control of illegal logging, etc.). Daily wages are most effective as incentives when they can be provided during times of low demand for agricultural labor, when the work to be provided is within the capacity of the villagers, when the work provides real benefits to the community, and when the work is on communal or nationally-owned land (such as a protected area). The work is usually best done by individuals acting as a community, in which case a community cooperative may be the best instrument for distributing the payments (thereby enhancing its influence in the community).

Direct Incentives In Kind

Direct incentives in kind normally work best in villages which are relatively poor and under-developed (which typically applies to the villages closest to national parks).

Food. An increasingly common incentive available for biodiversity projects is so-called "food for work," where villagers undertake community development work in exchange for food. This incentive works best in villages with seasonal under-employment, and where specific conservation-oriented activities can be undertaken. It provides an excellent motivation to work on conservation projects, especially if the community already has indigenous institutions which can recruit the labor, identify the priority tasks to be undertaken, supervise the work, and distribute the food. (See Case Study 10 for an example of how food for work has helped to improve the management of a protected area in China.)

Food for work programs usually depend on an international organization which provides the source of the food and influences where and how it is to be used. The major source of such food is the US Government, through Public Law 480, which provides food to the World Food Programme of the United Nations, and a wide range of development NGOs. The agreements usually need to be generated by a trained community development worker who is aware of the possibilities. The challenge is to link this aid to projects which enhance the sustainable management of biological resources.

Animals. Providing improved breeds of livestock to rural communities can be used as an incentive to increase productivity on high-quality pasture, thereby reducing pressure on marginal lands which are better left to wildlife. Governments or community development NGOs can also provide breeding stock of new species for domestication, such as cane rats *(Thryonomys)* in West Africa, capybara *(Hydrochoerus hydrochaeris)* in Brazil, or babirusa *(Babyrousa babyrussa)* in Indonesia; such domestication can contribute to conservation by producing improved forms (i.e., forms which meet human needs more efficiently than does the wild progenitor). Wild animals can also be provided as an incentive; in Brazil, for example, the government provides pairs of golden lion tamarin *(Leontopithecus rosalia)* for reintroduction into private forests, thereby encouraging land-owners to conserve their forests.

Access to resources in areas where sustainable harvesting is a management objective. In many countries, forests, wildlife, and fisheries have been nationalized or otherwise brought under central government control in the past few decades, primarily to facilitate higher levels of harvesting. Chapter 3 showed that this has often resulted in overexploitation and insufficient resources being available to local people for construction, firewood, minor forest products, and sources of protein. Case Study 7 (Sagarmatha National Park, Nepal) and Case Study 12 (forest management in the inner delta of the Niger, Mali) provide illustrations of this breakdown in traditional management systems, and suggest steps that can be taken to re-build the systems.

When legal restrictions (i.e., disincentives) are placed on certain species or areas, the local people often feel that a resource that was traditionally theirs has been removed from their control. Overnight, the village hunter has become a poacher. Once the hunter is outside the law, and is no longer hunting a resource which belongs to the village, the economic incentives to overharvest tend to overwhelm the incentives to conserve.

Further, if responsibility for managing the forest or fishery is moved from the nearby village to a distant forest department, fisheries service, or national park service, the traditional systems of incentives and disincentives begin to break down and over-exploitation of the resource becomes more common.

The usual approach to establishing and managing national parks and state forests has tended to make such resource conflicts inevitable. National parks, as defined by IUCN, are relatively large areas where "the highest competent authority of the country has taken steps to prevent or eliminate as soon as possible exploitation or occupation in the whole area and to enforce effectively the respect of ecological, geomorphological or aesthetic features which have led to its establishment" (IUCN, 1985). This reflects the view that areas of outstanding value to the nation as a whole can be managed to protect biological resources only if residents are kept out of the area.

Recognizing that this approach has caused serious difficulties for both local communities and the protected area authorities (which typically operate with insufficient manpower and budgets to carry out their assigned tasks), IUCN has designated a series of other types of protected areas where human exploitation is permitted on a sustainable basis (Box 8). While national parks need to be protected against human exploitation on a commercial scale, other types of protected areas—such as game reserves, biosphere reserves, protected landscapes, and multiple-use management areas—can be established around the strictly protected areas to prevent them from becoming biologically impoverished islands, or can stand by themselves to make important contributions to systems of land management. These categories of protected areas—notably IV to VIII in Box 8—can contain sustainable utilization of resources as a management objective, to both conserve biological diversity and provide sustainable benefit from the use of those resources. (See McNeely, Miller, and Thorsell, 1987, for more details on how the system of categories can contribute to the design and establishment of protected areas.)

Experience has demonstrated that **equitable access to resources can provide a powerful incentive for conservation, provided that this access is made possible within a structure of community or private responsibility for the continued productivity of the resource.** A number of the case studies (5-9, 11-16, 18) demonstrate that local people are well able to manage their own resources to provide their communities with a sustainable yield of goods and services; such controlled access to at least some of the resources contained within some categories of protected areas is often a necessary procondition for the conservation of biological resources in these areas (see also McNeely and Miller, 1984, and McNeely and Thorsell, 1986).

Even the strictly protected areas (generally Categories I to III) can provide some resources, through spill-over of game into so-called "buffer zones," where some restrictions on the harvesting of resources give an added layer of protection to the core area, while providing access to villagers for resources they require. Oldfield (1988) has provided detailed guidelines on the establishment and management of buffer zones, which can include protected area categories IV to VIII, plus selectively logged production forests, hunting areas, natural forests used by villagers for collecting firewood and other forest produce, forest plantations, perennial crops, and pastures or natural grazing areas.

In such areas, controlled hunting can be an important source of protein for local people, provided that the management authority is able to regulate which and how many animals are harvested. In the marine habitat, traditional forms of resource management have often proven effective, as illustrated in Case Study 14, on the coastal fisheries of Japan. Where traditional management systems have broken down, new management systems can ensure equitable access to resources and thereby provide incentives for conservation (see Case Study 9, on marine resources in Quintana Roo, Mexico, Case Study 16, on marine resources in the Philippines, and Case Study 25, on Caribou management in the Arctic).

Incentives can also be effective in protecting forest plantations. In Cianjur, Java, landless farmers are willing to protect teak and

Box 8: Categories and Management Objectives of
Protected Areas

While all protected areas control human occupancy or use
of resources to some extent, considerable latitude is available
in the degree of such control. The following categories are arranged in ascending order of degree of human use permitted
in the area.

1. *Scientific reserve/strict nature reserve.* To protect nature and
   maintain natural processes in an undisturbed state in
   order to have ecologically representative examples of
   the natural environment available for scientific study,
   environmental monitoring and education, and for the
   maintenance of genetic resources in a dynamic and
   evolutionary state.

2. *National park.* To protect relatively large natural and scenic areas of national or international significance for
   scientific, educational, and recreational use, under management by the highest competent authority of a nation.

3. *Natural monument/natural landmark.* To protect and preserve nationally significant natural features because of
   their special interest or unique characteristics.

4. *Managed nature reserve/wildlife sanctuary.* To ensure the natural conditions necessary to protect nationally significant species, groups of species, biotic communities, or
   physical features of the environment when these require
   specific human manipulation for their perpetuation.

5. *Protected landscapes.* To maintain nationally significant
   natural landscapes characteristic of the harmonious interaction of man and land while providing opportunities for public enjoyment through recreation and
   tourism within the normal life-style and economic activity of these areas.

6. *Resource reserve.* To protect the natural resources of the
   area for future use and prevent or contain development
   activities that could affect the resource pending the establishment of objectives based on appropriate knowledge and planning.

Box 8 (continued)

    7. *Natural biotic area/anthropological reserve.* To allow the way of life of societies living in harmony with the environment to continue undisturbed by modern technology.

    8. *Multiple-use management area/managed resource area.* To provide for the sustained production of water, timber, wildlife, pasture, and outdoor recreation, with the conservation of nature primarily oriented to the support of the economic activities (although specific zones can also be designed within these areas to achieve specific conservation objectives).

Source: IUCN, 1985

Shorea plantations in exchange for secure rights to the vegetable and fruit crops they inter-plant on this land. Perum Perhutani, the Indonesian State Timber Corporation, has entered into an agreement with local farmers that defines the rights and responsibilities of both partners in managing the concerned forest. Perum Perhutani's ownership of the land has not been weakened, but farmers have received a useful incentive—the secure right to products grown on forest lands—in exchange for a prescribed set of protection activities (Hamilton and Fox, 1987).

    In some cases, harvesting of some resources from even a national park can be a positive management step, while also providing an economic incentive for local people. In many parts of the tropics, rainy season growth of grass and canes can be far in excess of the needs of the protected fauna. In such cases, public access to these protected areas can earn income for the protected area, earn public support from the local people, and reduce the threat of fire during the dry season. Limited access to biological resources in certain parts of a reserve can be an important incentive for respecting restrictions on the exploitation of the resources in other parts of a reserve. Where traditional people live inside a reserve, they are often permitted to continue their traditional way of life, within restrictions established by the protected area authority. MacKinnon *et al.* (1986) describe a system of zones where traditional exploitation is permitted, including such activities as:

- fishing without poisons or explosives;
- traditional hunting of non-protected species without traps, modern weapons or use of fire;
- collection of gums and resins (provided trees are not killed in the process);
- gathering of wild fruits and honey (provided trees are not cut or burned);
- collection for own use of naturally fallen wood for lumber or fires;
- cutting of bamboo, reeds, thatch or rattan; and
- seasonal grazing of domestic animals, where native grazing species are not an important component of the park's resources.

Using public access to resources as an incentive needs to be balanced by disincentives against abuse of the resources. These disincentives need to be supported by the community, as illustrated in case studies 4, 6, 9, 11–16, 18, and 20; they are often most effective when traditional control structures are used to supplement laws and regulations promulgated by the central government.

As with incentives, disincentives controlling excessive use need to be applied with care. In Mali, for example, the Forest Code requires that farmers secure a free permit before harvesting trees considered to be of national economic importance (even when those trees occur on the farmer's own land). Failure to secure a permit can result in a fine levied by the Forest Agent, and since the permit is free of charge, the Forest Department feels that no farmer could reasonably object to securing one before harvesting a tree. However, securing a permit involves costs in kind (especially in terms of time and aggravation to seek the permit), and can lead to disputes with the Forest Agent on the extent of the proposed harvest. The farmer's perception of the value of his investment in trees may be reduced as a result of the permit system, making it more difficult for him to predict the returns associated with tree production.

The issue of controlling local public access to biological resources considered to be of national or international importance is perhaps the foundation of all conservation problems. The combination of appropriate incentives and disincentives, applied in conjunction with a system having a range of different levels of protection, such as that presented in Box 8, may be the best means of addressing public access problems.

INDIRECT INCENTIVES TO PROMOTE COMMUNITY SUPPORT FOR CONSERVATION

Less direct incentives may also be employed. The local people may be provided various forms of support to enable them to engage in profitable enterprises which derive from the protected area, including providing food and handicrafts for visitors, providing permanent labor to park management, and managing tourist facilities. These indirect incentives have the great value of providing villagers with the means to develop their own capacity to benefit from government efforts to conserve biological resources.

Indirect Fiscal Incentives

Tax incentives involve individuals or corporate entities being partially or totally exempted from government taxes (on land, income, sales, inheritance, or capital), in return for conservation-related behavior. The intention is to generate greater investments in conservation-related efforts, or to compensate farmers for other sacrifices. For example, farmers living near national parks may be offered tax exemptions for investing in appropriate activities in buffer zones established around a protected area, land tax exemptions (or deductions) may be provided for forest plantations (at least until they start producing harvestable products), or tax deductions can be provided for other investments which contribute to biological diversity, including nature tourism operated by village cooperatives, local investments in reforestation, and wildlife ranching.

Tax incentives are seldom very useful for small farmers who operate largely outside of the tax system, but they can have important implications for large land-owners, or for commercial operations. We have seen in the case of Brazil that such incentives can often have negative impacts on conservation, unless they are carefully designed to address objectives which lead to sustainable use of biological resources.

Security, guarantees, and insurance. Many rural people look upon forests and other natural ecosystems as reserves in times of particular stress, such as drought, famine, or domestic unrest. In order to reduce pressure on biological resources, which is likely to be particularly heavy in times of stress, governments can devise other mechanisms to provide the desired security. Such incentives might include food reserves, guaranteed access to necessary resources, and insurance against risks such as crop failure.

Indirect Service Incentives: Community Development and Biological Diversity

Local communities usually realize that sustainable utilization is to their benefit; but they are often forced by circumstances to harvest biological resources at a rate higher than can be sustained. In order to reduce their pressure on resources, alternative means of earning a living must be provided to these communities. Most rural communities require outside services if they are to improve their productivity and learn to manage their resources on a sustainable basis under modern conditions. Economic incentives are particularly effective when they support enhanced community development, linked explicitly with changes in behavior.

The intention of such indirect service incentives should be to ensure that the local people steadily reduce their dependence on outside inputs and build their self-reliance.

Community development activities which can be provided as part of an incentives package may include schools, health clinics, family planning, sanitation, community centers, electricity, roads, markets, and water systems. Paying for such incentives can be expensive, but costs can be reduced by the government (or other development agencies) providing expertise and materials, and the community providing labor. Such joint activities can help build community spirit, and provide a context for ensuring that the linkage between assistance and expected change in behavior is reinforced.

In some cases, community development activities are already being planned or implemented in communities which are important for conserving biological resources, in which case linkages with changed behavior toward conservation can be incorporated with little additional cost.

Agricultural inputs can also be provided, helping to rehabilitate soils and promote diverse and sustainable agro-ecosystems. Such inputs might include: forest plants for timber and fuelwood; perennial crop species, particularly fruit trees; saplings and seeds of fodder species; improved seeds; agricultural chemicals; fencing; breeding stock; storage facilities; and tools or other equipment.

Improved use of resources can signficantly reduce human impact on natural ecosystems. For example, incentives which promote reduced reliance on firewood collected from the surrounding countryside—such as improved stove designs, firewood plantations, electricity, and subsidized prices for kerosene—can often have a major positive impact on the forest.

Rural development projects that make use of native biological diversity can be effective in demonstrating significant uses of biological resources and providing benefits to local communities; this provides rural people with an incentive to conserve their local biological resources because of the increased economic return they obtain from it through their own efforts. These can include development and marketing of foodstuffs, medicines, arts, crafts, and other products from native plants and animals (though sustainable use will require developing commensurate management mechanisms); domestication of wild resource species of both plants and animals; and development of nature-based tourism in a manner compatible with the social and cultural values of the community (Prescott-Allen, 1986).

Education and training are often considered by rural people to be the most useful outside contribution they can receive. The desire for well-educated children is virtually universal, and provision of schools is a powerful incentive; it also provides a ready medium for promoting the conservation message, thereby helping to ensure the sustainability of the investment in improved management of biological resources. For adults, training programs can often be provided as part of larger development efforts and linked with other incentives. Providing training for cooperatives managers, local health workers, marketing specialists, and so on, is warmly welcomed by most communities. Education should become institutionalized as soon as possible, so that children are made aware of the very specific values attached to their surroundings; this is particularly true in the case of national parks and other categories of protected areas. However, in societies where awareness of the linkages between resource management and human welfare is low, such education activities should be seen as a long-term investment.

Social Incentives for Conserving Biological Resources

As suggested above, local people who live in or near a protected area, or who have major contributions to make to national objectives for sustainable use of biological resources, should receive high priority for development activities which reduce their dependence on the unsustainable exploitation of goods provided by these resources and to enhance the sustainable flow of environmental services.

Community organization. **The establishment of strong village-level institutions can be the single most effective incentive for behavior which contributes to conservation of biological resources,** and case studies 3–7, 9, 11, 12, 14–16, 18, and 20 address this issue in more detail. In many cases, such institutions already exist and merely need to be strengthened or reinstated. In the Khumbu region of Nepal, for example, the Sherpa people have a system of *shing-i-nawa*, or forest guards, where several men from a village are elected to protect the forest which provides wood for construction and fuel, and protects the village from landslides and avalanches. They have the power to prevent cutting of protection forests, determine where trees may be cut, inspect firewood stocks in people's houses, and levy appropriate fines for transgressions. Their power is reinforced by annual celebrations where the fines are paid and the perpetrators are subjected to good-natured ridicule by their peers.

One of the most interesting examples of a totally new institution has been the establishment in Thailand of village level "Environment Protection Societies" (EPS), which integrate social development with conservation activities adjacent to Khao Yai National Park (Case Study 4). This link was immediately and forcefully made by requiring that villagers pledge to cease poaching and encroachment as a prerequisite to EPS membership. In addition, conservation themes were woven into all project activities and programs. Practical conservation training, including soil conservation, tree planting and other skills, has been an integral project component. Conservation awareness programs stressing links with development have provided new perspectives and new options to villagers.

The EPS was considered necessary because the villages around Khao Yai are both relatively new and relatively poor, with weak community institutions. This crucial first step enabled subsequent project activities to build on the new foundation of social cohesiveness. Once the EPS was established, villagers themselves made all decisions regarding project activities, with guidance from professionals.

Land tenure governs the use and disposal of land and its products so that the use of the land can be stabilized. When villagers do not have secure title to their land, they have little incentive to make investments that would ensure sustainable use; credit is more difficult to obtain when tenure is insecure; and insecure tenure may bias the choice of crops against perennials, tree crops and forest plantations which tend to be environmentally more benign than annual field crops. Villagers lacking secure tenure are therefore forced to continually clear new land, often destroying forest of high biological value and leaving little but wasteland behind.

On the other hand, farmers with secure tenure have a strong incentive to make investments in ensuring the long-term productivity of the land, since they will be able to reap the benefits from the investments they have sown. Therefore, clarifying land tenure is often the first step in obtaining any sort of government grant or subsidy, and may require governments to take a somewhat flexible attitude in the light of land-use legislation that may date from colonial times. Further, granting tenure for agricultural lands must be linked with controls on the uses of other land, particularly in legally-protected areas of outstanding importance to the nation or the larger community.

It is important to note that secure land tenure can include common property or collective local decision-making as well as private property. In parts of Papua New Guinea, for example, the land is owned by deceased ancestors and this provides excellent protection against alienation of the land rights from those who have usufruct.

Employment. Directly and indirectly, a protected area can significantly enhance employment opportunities in the region. In Rwanda, for example, the Parc National des Volcans (13,000 ha) conserves a population of mountain gorillas *(Gorilla gorilla beringei)* which in turn is the country's major tourist attraction. The guides and porters who take tourists to see the gorillas are former poachers who now earn cash from conserving the animals.

A certain number of local people may be employed directly by the management authority, or in catering for visitors to the area or providing ancillary services. Whenever possible, local people should be employed as reserve staff or as concessionaires in preference to outsiders from more distant towns. This keeps locally generated wealth within the communities immediately adjacent to the protected area. In some cases, such as the Galapagos Islands, a park or reserve can stimulate the whole local economy, especially if the monies deriving from the reserve and its visitors are used and circulated within the region. Paid employment may be less easily recognized as an incentive by the local communities, however, because payment for labor is not always clearly related to the protected status of the area. It is valuable to underline the relationship between

Box 9: The Capacity of Communities to Benefit from
Incentive Schemes

The capacity of any given village or community to benefit
from incentives will vary considerably from community to community. The effectiveness of a package of incentives aimed at
a specific community depends on a number of factors, including:
- the major *objectives* of the incentives scheme (the most important issue here is to be very clear and explicit about
  what conservation objective is to be achieved by the
  incentive);
- the *capacity* of the community to absorb incentives (villages
  with well-developed institutions will usually be able to absorb incentives more effectively than poorly organized villages, which may first require the development of
  appropriate institutions);
- the initial *state of the biological resources* to be managed (incentives to manage existing resources are different from incentives to rehabilitate resources that have been depleted);
- the *level of motivation* of the community (communities which
  are eager to cooperate and take advantage of opportunities such as tourism are quite different from communities which need to be convinced that cooperation is in their
  own best interest; in the latter case, an initial promotion
  campaign may be required);
- the *constraints* which the incentives are intended to overcome (these can include: lack of title to land; unclear
  responsibility for biological resources to be conserved; insufficient information about available options or rights
  under the law; lack of access to resources, expertise, or
  appropriate markets; and insufficient awareness of the
  benefits available from conservation action);
- the *effect of time* on the incentives (including the time required to apply the incentive, the time over which the incentive needs to be applied, the time required for the
  incentive to bring about the desired change in behavior,
  and the time to recover any recoverable investments); and
- the *method of distributing* the incentive to the community
  (communities with strong institutions may use them to distribute the incentives, while other mechanisms may be required in other cases; this will obviously vary with the
  objectives and degree of motivation).

regional employment and other benefits through the protected area extension and information program.

In Costa Rica's Guanacaste National Park, loss of land tenure has been combined with employment to provide explicit incentives for conservation. As of late 1987, six managers of local ranches purchased to become part of the park had been hired as park site managers. They were allowed to stay in their houses and cultivate a small portion of their fields to supplement their salaries. They enlarged their houses into biological stations, mapped local vegetation, and received training in public relations and handling of poachers. The new biological stations provide food and shelter to visiting scientists, which also brings cash income to the former managers (Allen, 1988).

Information. It is apparent that market efficiency is best achieved when all parties involved are fully informed of the benefits and costs involved in a particular market decision. Therefore, information is often an effective incentive, as it can be the best molder (and reflector) of community standards. When villagers become aware of what incentives are available, they are better able to take advantage of them. When they are aware of the long-term consequences of their actions, they are more likely to behave in ways that promote sustainable use (provided other imperatives enable them to do so). When they are aware that the biological resources they are conserving are of national or international importance, they are more likely to take pride in their conservation activities. In short, villagers need to be fully informed about the benefits of using biological resources sustainably, and about the government incentives which are available for assisting them to do so.

CONCLUSION

Which members of a population have their access to biological resources enhanced and which members have it restricted by government policies is of profound importance in determining whether the resources will make a sustainable contribution to society. People living in and around the forests, wetlands, and coastal zones are of paramount importance to sustainable use of biological resources. They, rather than governments, often exercise the real power over the use of the biological resources. They should be given incentives to manage these resources sustainably at their own cost and for their own benefit.

Well-designed packages of economic incentives can ensure that the local communities which are most directly affected by both conservation and by over-exploitation can earn appropriate benefits from behavior which is in the national interest.

Ideally, such packages should consist of the following elements:

- Establishing *what are the biological resources* for which management needs to be enhanced.
- *Estimating the economic values* of these resources.
- Establishing *conservation objectives* for the package of incentives and disincentives.
- *Determing perverse incentives,* the national social and economic policies that have encouraged the community to over-exploit biological resources.
- *Collecting information about the community,* including determining what biological resources the community is currently using, how the resources are being managed by the community, the degree of awareness about controlling regulation, and possible alternative sources of income.
- *Designing specific packages of incentives* to meet the highest priority needs of the villagers, and ensuring that the incentives package is linked with other development activities.
- Establishing a *structure of responsibility* for the biological resources in the area, often through the use of village-level institutions.
- Incorporating *packages of disincentives,* through legislation, regulation, taxation, peer pressure, and appropriate levels of penalties.
- Providing appropriate *information and public education* to the target audiences on both incentives and disincentives.
- Establishing a means of *monitoring and feed-back,* so that necessary changes can be instituted as the incentives system adapts to changes.

The intention of all packages of incentives and disincentives aimed at the local community should be to ensure that the local people steadily enhance their self-reliance and self-esteem, and reduce their dependence on outside inputs. The effectiveness of such systems, however, depends on supporting policies from local and national government.

CHAPTER FIVE

THE USE OF INCENTIVES AT THE NATIONAL LEVEL

INTRODUCTION

While community self-reliance is a worthwhile goal, all communities are part of the larger nation, and the biological resources which support the community are also of considerable interest to the nation and the world. Further, incentives at the local community level are likely to require considerable support from compatible policies at the national level. Finally, governments seldom have sufficient capital or labor to manage their nation's biological resources in an optimum way. Packages of incentives and disincentives which bring in additional funds for conserving biological resources can be an essential means of implementing national development goals.

While each government will need to determine its own objectives for conserving biological resources, the following general objectives are broadly relevant for supporting systems of incentives:

- To coordinate economic and development policies so that appropriate incentives packages can be implemented at the community level.
- To provide the policy basis by which the various sectors having potential or actual positive or negative effects on biological resources are made aware of how they can contribute to conservation.
- To enable the resource management sectors to develop supplementary sources of funding, thereby earning sufficient funds to carry out their mandates effectively.

- To provide a framework within which the private sector can support national objectives for conservation of biological resources.

An important point to keep in mind is that biological resources do not occur only in protected areas, and economic incentives may also be used more generally throughout the country to encourage settlement patterns, plot sizes, and productive systems that are directed at the sustainable use of the resources of forest, wetland, and sea. These incentives should accompany regulations on matters such as mesh size of fishing nets, rate of clearing of forest for agricultural use, and harvest and marketing of wild plants and animals.

NATIONAL GOVERNMENT POLICIES WHICH ENABLE ECONOMIC INCENTIVES TO FUNCTION AT THE COMMUNITY LEVEL

The incentives which may be required at the community level usually require commensurate policies at the national level. If villages located around national parks are to receive special attention from national and international development agencies, for example, a directive to this effect needs to come from the relevant government authorities. The specific policies required at the national level will derive from what is required at the community level to implement national policies for conserving biological resources. The following points will be generally useful.

Direct Incentives in Cash

Capital is a scarce resource, and direct cash incentives provided in support of conserving biological resources could instead have been used in a number of other ways. Therefore, the government policy on which direct cash incentives can be used will need to consider competing demands, drawing on the concept of Marginal Opportunity Cost. Viable cash incentives will need to be profitable, so that the economic benefits exceed the economic costs by greater amounts than the benefits offered by alternative uses.

Further, issues of equity will need to be considered. What is a fair direct return to villagers, and how much of the rent collected from exploiting a biological resource should go to the government (perhaps for providing indirect returns to the community)? What mix of recoverable cash incentives (e.g., loans, revolving funds) and non-recoverable cash incentives (e.g., rewards, grants, and fees) is most appropriate?

Direct Incentives in Kind

The full development of direct incentives in kind requires a significant government effort aimed at bringing together the community which requires the incentive and the agency which can provide it. This is particularly sensitive, in that the communities which most need such support are often least able to negotiate on their own behalf. Such communities may need a supportive development NGO to assist them in identifying their needs, and ensuring that these needs are transmitted to the right quarters (see Case Study 4).

Development of buffer zones, game management areas, and other lands which are designed to provide sustainable benefits to local people often requires coordination among a range of agencies, including the resource management agency (whose authority may end at the boundary of a forest or national park, or at the shoreline) and the various development-oriented agencies (agriculture, irrigation, tourism, transportation, etc.). Providing access to biological resources by local people implies an opportunity cost, so appropriate determinations of Marginal Opportunity Cost can be a useful tool here.

Public access to protected resources is often the most difficult issue, because the relevant legislation has often been modelled on temperate-zone legal, social, and ecological situations which are no longer relevant in the tropics. The challenge here is to find the right balance between exploitation, sustainable use, and degradation of the biological resource; this is often a value judgement, depending very much on the management objectives which have been established for an area or a resource. In many cases, legislation may need to be reviewed in light of current conditions and based on an accurate estimation of costs and benefits.

A useful tool here is adopting a series of management categories for species and areas, with different categories corresponding to different management objectives (as described in Box 8).

Indirect Fiscal Incentives

As suggested in Chapter 3, indirect fiscal incentives can perversely cause great damage to biological resources. But properly designed, they can also support conservation. Government policy needs to compensate for externalities and other market failures so that each activity which affects biological resources is fully accountable for its social and biological costs. Economic incentives in the form of taxes on activities that generate negative externalities (social costs) and subsidies on activities that generate positive externalities (social benefits) should be considered.

Numerous indirect fiscal incentives are available to governments for supporting conservation of biological resources:

- The use of state and federal tax exemptions and credits is an important mechanism for encouraging protection of important natural lands held by the private sector. These programs generally operate by providing economic advantages to persons or corporations selling or donating such areas to various agencies. The incentives can be given in the form of personal or corporate income tax deductions, involuntary conversions, land exchanges, reduced state and local property taxes, and favorable valuation for estate taxes.
- Tax deductions for conservation purposes are very common in industrialized countries, especially for donations to conservation NGOs; such incentives provide one of the best ways of expressing existence value, encouraging various private sources to contribute to biodiversity projects.
- Import taxes and duties can be waived for equipment which is required for activities supporting conservation of biological resources.
- Governments can offer increased depreciation or other tax-deduction schemes for investments which contribute to biological diversity, such as appropriate tourism developments based on the resources of national parks or other protected areas.
- Taxes can be used both to raise revenue and to provide an economic disincentive against engaging in environmentally destructive behavior, such as certain forms of land development.

The special case of private land. In some countries, much of the remaining land covered in natural vegetation remains in private ownership. In such cases, governments can help protect these areas by encouraging the private sector to take affirmative responsibility for conserving them, by offering substantive economic incentives such as tax credits, subsidies, etc. Subsidy programs financed by governments can directly compensate private landowners for not converting or destroying areas of outstanding importance. The protection subsidies provide to these areas is effective only to the extent that compensation for protecting the area exceeds competing economic uses for the land. Such incentives for conservation need to be accompanied by the removal of perverse incentives to alter such areas.

Indirect Service Incentives

The service incentives outlined in Chapter 4 in general link community development with conservation of biological resources. Most governments have extensive community development programs, often supported by a multiplicity of development agencies. The main policy shift required here is to give higher priority to the communities which are closest to protected areas or other areas of outstanding value for the biological resources they contain; this requires that the various development agencies be made aware of the special value of biological resources, and their relevance to sustainable development.

In addition, government research and development programs could provide important assistance and stimulus to community-level development of native biological resources, through product development, market research, evaluating potential new domesticates, training, research grants, and guaranteed development loans.

Indirect Social Incentives

As with the service incentives, the social incentives may require little more policy support than providing higher priority to key communities identified as being of particular importance for national conservation objectives. On the other hand, many communities will require support from central governments or national NGOs in building up the community organizations which are required to take advantage of the available incentives. This is particularly the case where villages are new, have a mix of ethnic groups, or are on the frontier; such communities may be inherently unstable, requiring outside influence to provide the social stability necessary for sustainable development.

COORDINATING CONSERVATION INCENTIVES: THE CROSS-SECTORAL APPROACH

It can be seen that taking full advantage of the opportunities for using economic incentives to conserve biological resources in support of sustainable development requires coordination among a number of policies and levels. The World Commission on Environment and Development has pointed out that environment and development are not separate challenges, but are inexorably linked. WCED (1987) states that, "Development cannot subsist upon a deteriorating environmental resource base; the environment cannot be protected when growth leaves out of account the costs of environmental destruction. These problems cannot be treated separately by fragmented institutions and policies. They are linked in a complex system of cause and effect."

Many of the problems in conserving biological resources are related to the fact that responsibilities are divided into sectoral units, leading to fragmentation, poor coordination, conflicting directives, and waste of human and financial resources. This can only be overcome by integration, by examining the impact of decisions in one sector on the ability of another sector to depend on the same resources. Integration is not easy, and in some respects it is not very practical. Still, an optimal balance point can be found where the benefit of considering secondary impacts (or externalities) is overtaken by the cost of doing so; in most cases, this balance point lies well beyond the current practice of taking decisions based on a very narrow range of sectoral considerations.

The role of conservation strategies

One means of initiating improved policy coordination is through preparing a national conservation strategy (NCS) or a subnational conservation strategy (SNCS). In one form or another, the NCS process has been initiated in some 35 countries. Focussing on national planning and the range of decisions taken by the public sector on the use of biological resources (either deliberately or by default), an NCS can address many of the most fundamental policy issues faced by governments seeking to use their biological resources on a sustainable basis.

The first requirement for a successful NCS is the participation of the widest possible range of actors in defining the issues and identifying possible courses for action. No matter how broadly based a government may be, the nature of the public sector (or indeed of any centralization of power) limits the range of issues which can effectively be considered. The NCS process places government in partnership with NGOs, citizens' groups, universities, industry, financial institutions, and many others in seeking to relate the use of biological resources to national development objectives. It therefore provides an important (and generally non-threatening) forum for reaching national consensus on the use of biological resources. Few better mechanisms seem to exist.

Some cross-sectoral linkages

Several tools have been developed to incorporate what once were regarded as external considerations in development decisions. Environmental Impact Assessment (EIA) is one such tool, and its application has yielded many benefits. Yet EIA generally only offers guidance once fundamental choices among available options have been taken. The NCS approach, in developing a framework where environmental concerns can be related to development objectives, offers the possiblity to approach a more appropriate balance point through a process of consensus-seeking.

The remainder of this section discusses some of the cross-sectoral linkages that affect decisions on the use of biological resources at the national level, and how incentives can be used in these sectors to promote more effective decision-making in the use of biological resources. In most countries, the primary responsibility for conserving biological resources rests with conservation agencies, such as departments of national parks, wildlife management, forest protection, soil conservation, watershed management, and fisheries. However, many other sectors also have powerful impacts on biological resources; to date, negative impacts—often through external effects and perverse incentives—have tended to outweigh positive impacts. But incentives are available to a wide range of government agencies, NGOs, and private enterprise for contributing to national objectives for conserving biological resources.

Perhaps more important is the observation that the real resource managers are the rural people, especially women, who deal with resource limitations on a day-to-day basis. But the ministries of finance, foreign affairs, commerce, and trade set the policy framework within which individuals operate. The WCED stressed the importance of de-centralizing the design and implementation of resource

Box 10: Policies to Promote Integrated Action

The close link between rural development and conservation of biological resources demonstrates that action in either alone will not solve the problem. Instead, conservation needs to be woven together with agriculture, forestry, fisheries, transport, national defense, and other efforts. The following major policy components might be included in such integrated action.

- The many economic and financial benefits of integrated rural development linked with conservation of biological resources need to be quantified and brought to the attention of policy makers.
- Conflicts between the various activities of agriculture, fisheries, forestry, conservation and rehabilitation need to be identified in integrated plans and programs.
- Institutional reform and improvement may be require as part of good design and implementation of integrated sectoral development plans and programs.
- Legislation may need to be formulated consonant with the socio-economic patterns of the target group and the natural resource needs, both to institute disincentives and to ensure that incentives carry the power of law.
- Policies and legislation in other sectors need to be reviewed for possible application to conservation of biological resources and community involvement in such work.
- Effective incentives need to be devised to accelerate integrated development to close any gap between what the individual sees as an investment benefit and what the government considers to be in the national interest.
- The rural population needs to be involved in the design and follow-up of plans and projects, not simply their implementation (Velozo, 1987).

management programs so that project activities are sensitive to local conditions; a major challenge is the proclivity of central government agencies to hold on to control of the resources they are intended to manage.

Water resources development can often provide effective incentives for conserving biological resources. MacKinnon *et al.* (1986) and

McNeely (1987) have provided examples demonstrating that watershed protection has helped justify many valuable reserves which otherwise might not have been established, and shown that irrigation and energy agencies can make powerful potential allies for protected areas which protect watersheds.

Some water development projects may require large scale infrastructure development that threaten biological resources. Even when large dams cause considerable negative environmental impacts in downstream areas, they often also provide a direct incentive for conserving steep-sloped, erosion prone and high-rainfall upper watersheds which serve as "water factories" for downstream developments (Barborak, 1988b).

Other such projects, however, particularly those designed to meet local water, irrigation, and power needs, require minimal habitat alternation yet provide an economic rationale for total or partial protection of catchment basins. In Honduras, for example, the La Tigra National Park, a 7500 ha area consisting mainly of cloud forest, produces a high quality, well-regulated water flow throughout the year, producing over 40 percent of the water supply to Tegucigalpa (the capital city). Some 25 small collection facilities scattered throughout the park require only limited maintenance because the water is so pure and free of sediments. Because of its value for watershed protection, La Tigra is the focus of a major investment program involving a series of economic incentives for villagers living in the buffer zones (see Case Study 13).

In many cases, the total costs of establishing and managing reserves which protect catchment areas can be met and justified as part of the hydrological investment. Hufschmidt and Srivardhana (1986) have shown that an annual expenditure for watershed protection related to the Nam Pong Reservoir in northeastern Thailand of about $1.5 million per year would be justified in terms of benefits to the reservoir.

MacKinnon (1983) examined the condition of the water catchments of 11 irrigation projects in Indonesia for which development loans were being requested from the World Bank. The condition of the catchments varied from an almost pristine state to areas of heavy disturbance due to deforestation, logging or casual settlements. By using standard costings for the development of the protected areas, reforestation where necessary, and any settlement of families required, the costs of providing adequate protection for the catchments were estimated. These ranged from less than 1 per cent of the development costs of the individual irrigation project in cases where the catchment was more or less intact to 5 percent where extensive reforestation was needed, and a maximum of about 10 percent of development costs in cases where resettlement and reforestation were required. Overall these costs were trivial compared to the estimated 30 per cent to 40 per cent drop in efficiency of the irrigation systems expected if catchments were not properly safeguarded.

It is evident that the costs for protecting watersheds should be an automatic component of irrigation projects, based on a sound foundation of ecological science and economic justification.

In addition, economic incentives in the form of water pricing or allocation of water rights would clearly improve efficiency and equity of water use as well as generate funds for maintenance of the irrigation system and protection and management of the watershed, with additional environmental benefits in terms of conservation of tropical forests and endangered species (Panayotou, 1987).

Public works. To maintain a variety of services to surrounding human settlements, public works and other departments may need to establish installations within protected areas or other lands of high value for conserving biological resources (MacKinnon *et al.*, 1986). Examples include:
- roads, canals, railways, paths crossing reserves;
- water pipes, oil and gas pipes, power lines, telephone cables;
- water stations, sewage works;
- hydroelectric dams, geothermal plants;
- telecommunications stations;
- meteorological stations, astronomical observatories;
- quarries or gravel pits.

In most cases, biological resources are better off without such installations. But when the national interest dictates that they be accommodated within a protected area, they should be expected to contribute financial support in the form of compensation to the affected area. At the Monte Verde Cloud Forest Reserve in Costa Rica, for example, an annual rental fee is paid to the park for the use of one of its mountain tops for telecommunications structures, providing a supplemental source of income which is used for park management programs.

Agriculture. Since tropical forests often grow on soils which are poor in cations or suffer from other deficiencies, attempts at agriculture are often followed rather quickly by abandoned fields and degraded vegetation. As a result, large areas of the tropics are covered by devastated landscapes which are productive only for grazing at a very low stocking density. With proper economic incentives, such areas can be made productive again, either for agriculture, forestry, or conservation of biological diversity. It may be more expensive in the short run to reconstitute damaged ecosystems than to conserve new lands, but it will often be as economically efficient in the long run to rebuild degraded local ecosystems rather than to exploit (and degrade, requiring rebuilding) other more remote land, which may itself be more sensitive to degradation and intrinsically less profitable (BOSTID, 1985).

The use of agricultural chemicals has often been subsidized, leading to inappropriate uses which have threatened biological diversity in many parts of the tropics. Such inappropriate incentives to promote consumption need to be analyzed relative to other sectors; better incentive systems might instead promote organic fertilizers and soil conservation.

Linkages between conservation and agriculture are also important in industrialized countries. Under a recent regulation adopted by the European Community, EC Governments may define certain areas of the farmed countryside as "Environmentally Sensitive Areas." Such areas are important in environmental terms, and their continued environmental protection depends upon the survival of the traditional forms of farming which give rise to their environmental qualities. Within ESAs, farmers are paid special grants in order to persuade them to continue to farm in a traditional way; they lose the payments if they switch to a more intensive form of production. ESA payments, therefore, can involve limitations on the amount of fertilizer which can be used, restrictions on changes of agricultural land use (such as from grazing to cereals), and controls over the dates at which meadows are cut for hay; they may also include positive payments to encourage practical conservation, such as woodland management or the restoration of archeological features. In the United Kingdom, some 400,000 ha are now covered by ESA designation, with the funds—currently some $18 million annually—coming from the agricultural budget.

A group of US-based NGOs called the "Committee on Agricultural Sustainability for Developing Countries" (CASDC) has suggested a series of criteria for developing sustainable farming systems.

Such systems are required if pressures on marginal agricultural lands are to be reduced, thereby enabling such lands to be devoted to conserving biological resources. Sustainable farming systems incorporate the following characteristics:

1. They maintain and improve soil productivity, quality, and tilth.
2. They augment the potential for achieving the highest possible efficiency in the use and conservation of basic farm resources (soil, water, sunlight, energy, and farmers' time).
3. They incorporate as much biological interaction as possible, including such processes as mulching, the use of nitrogen-fixing plants, the use of agroforestry techniques, and the use of intercropping and crop rotations to control pests and weeds.
4. They minimize the use of external inputs which endanger human health and damage the environment (some chemical fertilizers; non-selective pesticides and herbicides; and some forms of energy) and, instead, maximize the use of available, affordable, renewable, and environmentally benign inputs.
5. They avoid the contamination of groundwater by using only those fertilizers, pesticides and herbicides that do not penetrate below the plants' growing zone and then only in controlled doses.
6. They meet the needs of farm families for energy to work their land, cook, and heat from readily available and affordable energy sources.
7. They meet the needs of farm families for cash income, including from off-farm sources.
8. They are adaptive, so that, even as society evolves and communities change, they will strengthen communal cooperation, protect rural survival systems, through community support and sharing allow farm families to keep going in difficult times (famine, drought, and natural or political disasters), and make possible effective local management of community-controlled common property resources (ponds, woodlots, grazing lands, irrigation systems) in ways that permit equitable sharing of benefits.

In order for biological resources to be conserved in non-agricultural lands, these tests of sustainability must be developed and applied to all kinds of farming systems, from the intensive mono-cropping systems to animal husbandry to agroforestry to the vast numbers of mixed systems used by a billion small farmers (CASDC, 1987). Therefore, research organizations, development agencies, and governments need to support work on the continuing evolution of the concepts and practices of sustainability, provide encouragement and incentives for the adoption of sustainable agricultural systems (many of which were discussed in Chapter 4), and ensure that farmers receive their fair share of the benefits from conserving biological resources.

Primary production. As has been suggested earlier, considerable potential exists for increasing the economic returns to local communities from its harvests of native biological resources through development of the capacity to make and sell products from the harvested species, and by increasing the sustainable yield and improving quality through domestication of wild resource species.

In order to improve yield and quality, and to reduce pressure on species which are under significant pressure for harvesting (such as crocodiles, sea turtles, and antelope), many nations have begun wildlife farming or ranching schemes. Some such schemes involve taking eggs or young from the wild and rearing them in captivity, thereby ensuring a more reliable harvest and a higher survival rate; the intention is to provide significant economic benefits while reducing the pressure on the wild population. More significant are genuine efforts at domestication, which have far greater potential for reducing pressure on their wild conspecifics by providing an improved product; these projects are better considered as "new livestock production." Many, even most, of these schemes require tax incentives such as deductions of investments in infrastructure, tax holidays on any goods exported, etc., in order to become established and profitable.

Game farms can seldom drive out poachers of wildlife in national parks simply by underpricing them, because the species in the park are subsidized by the government and free for the picking. Disincentives, such as fines and jail sentences, are usually required as well. In addition, governments may also consider providing subsidies of a scale of magnitude such that the game farms can underprice poachers. But as the program succeeds in making poaching uneconomic and causing the species to become more numerous, poaching costs will fall (with rising population density).

Thus, the subsidy will have to increase until a viable population (net of poaching) is obtained, if the objective is to use game farms to reduce commercial poaching.

The government subsidy for game farms must include training, marketing, capital investment, and operating subsidies, if such enterprises are to become a rational alternative to collecting in the wild.

Tourism. Natural areas—mountains, rivers, wetlands, forests, savannas, coral reefs, deserts, beaches—are major attractions for tourists. Tourism can bring numerous socio-economic benefits to a country, in terms of creating local employment, stimulating local economies, generating foreign exchange, stimulating improvements to local transportation infrastructure, and creating recreational facilities. Positive effects on the environment often derive from these socio-economic benefits. Such positive effects may include:

- encouraging productive use for conservation objectives of lands which are marginal for agriculture, thereby enabling large tracts of land to remain covered in natural vegetation;
- promoting conservation action by convincing government officials and the general public of the importance of natural areas for generating income from tourism; and
- stimulating investments in infrastructure and effective management of natural areas.

These benefits can provide incentives for effective management of the natural areas which are tourist destinations, which in turn enhances the quality of the natural resources that attract tourists. Properly planned and managed tourism in natural areas is both non-polluting and renewable, and numerous examples exist where tourism has provided powerful incentives for conserving biological resources. Outstanding examples include Royal Chitwan National Park in Nepal, where tourism developments have been kept within rigorous limits and tourism has been a major justification for saving the endangered Great Indian Rhinoceros; and Tai Island in Fiji, where as a result of protection, subsistence fish catches have increased, tourist activity has expanded and the holders of traditional fishing rights are involved in resort management and boat hire.

However, biological resources can also be damaged by inappropriate tourism developments. McNeely and Thorsell (1987) have outlined the positive and negative impacts that tourism can have on such resources and recommend that the guiding principle for tourism development in natural areas should be to manage the natural and human resources so as to maximize visitor enjoyment while minimizing negative impacts of tourism development.

Four general principles are relevant for linking investments in tourism with conservation of biological resources:

- Planning for tourism development must be integrated with other planning efforts, particularly in national parks and other natural areas which are potential tourist destinations.
- Tourism authorities working with protected area managers should determine the level of visitor use an area can accommodate with high levels of satisfaction for visitors and few negative impacts on the environment, and ensure that this level is not exceeded.
- National policy should require environmental impact assessments (EIA) for all tourism development projects or programs, and specify the ways and means that the tourism development can provide economic benefits to both the local people and the natural areas which are the primary tourist destinations.
- For each major tourist destination based on the attractions of biological diversity, a management plan should be developed to specify objectives for both tourism and resource management, and to determine how sufficient income from tourism can be provided to the natural area to provide an incentive for improved management.

In short, tourism and conservation of biological resources can be natural partners, and each can benefit from the other if both are properly managed. Sufficient resources must be devoted to managing the natural areas, but it is often difficult to convince the governments who are responsible for budgets to allocate sufficient funds for this purpose. It is in the interest of both tourism and conservation that governments be so convinced.

Research. Research and information is urgently needed to examine cross-sectoral impacts of government policies on biological resources. Funding should be provided to national research bodies, relevant ministries, or universities as an incentive to encourage increased field research on conservation of biological resources. Relevant subjects for research could include:

- social science relating to environmental problems;
- the impact of reduction or elimination of biological resources upon the quality of life of the individual as well as the culture of the community;

- the effects of subsidies for agriculture and energy;
- the effects of price policies in general on the management of biological resources (going beyond subsidies for agriculture and energy);
- evaluation of the extent to which the positive impact of increased project funding for efforts to conserve biological diversity in developing countries has been offset by governmental austerity measures imposed by these same lenders;
- the sustainability of export diversification policies;
- the economics of forestry policy, and of river basin management;
- development of integrated farming systems that are not dependent on subsidies;
- detailed study of *all* economic values of an area.

In research on agronomy, agroforestry, and reforestation, unconventional and unorthodox solutions should be encouraged through economic incentives for discovering new solutions to environmental problems, especially those based on the use of native species.

Incorporating a research component into a development project is often both necessary for the project's successful implementation (e.g., surveys, assessments, monitoring and technology development) and a useful way of giving researchers practical field experience. However, such research should not end with the project, and means should be developed for ensuring that continuing research and monitoring becomes institutionalized as part of the development process.

ECONOMIC INCENTIVES AND RESOURCE MANAGEMENT AGENCIES

The main government reaction to the destruction of nature has been to establish national parks and other categories of protected areas, and today some 4 percent of the earth's surface is so protected. But despite major investments in protection, few protected areas anywhere in the world are managed effectively enough to ensure the survival in perpetuity of the biological resources they contain.

The government agencies with primary responsibility for conserving biological resources are under increasing stress, as human populations increase their pressure on biological resources at the same time that financial resources available for management are shrinking. Government resource management agencies therefore need to seek innovative ways and means of using economic incentives for attaining their objectives in ways other than direct budgetary allocations (see Chapter 7 for more details).

Part of the problem is that many protected areas could yield products in addition to environmental services, if they were managed in different ways. If alternative management policies could generate a stream of benefits from the protected area, in addition to environmental services, then the total benefits of the protected area will be increased, and this in turn will raise the opportunity cost of developing the area for other uses.

The guidelines presented in Chapter 8 describe how resource management agencies can build their capacity to implement incentive programs, but several additional points can be usefully addressed here, involving the use of incentives by resource management agencies to carry out their mandates more effectively.

Incentives for Staff

Virtually all governments suffer from a gross imbalance between the means devoted to enforcing conservation policies and the market value of the resources which are being protected; one consequence of this is that the salaries of officials assigned to enforce conservation measures are often exceedingly low in comparison to the worth of elephants, vicuna, tigers, and trees. While significant increases in salaries are probably not feasible, a series of other economic incentives could help ensure that conservation officials are rewarded at appropriate levels. Depending on local conditions, these could include:
- free or subsidized housing;
- special schooling allowances for work in remote areas;
- adequate equipment;
- cash awards for outstanding service;
- regular study tours to other countries in the region;
- public recognition for effective work;
- career development incentives which include requirements for field time;
- receiving technical information.

In addition, incentives are often required to encourage personnel to actually get into the field where much of the grass-roots work in conserving biological resources is carried out. One possibility is to provide a cash incentive in the form of an honorarium for staff when they are carrying out field work. The possibilities for the inappropriate use of such incentives are apparent, and particular care needs to be taken in their administration.

Community Service Voluntary Labor

Providing opportunities to local people, students, visitors, or private firms for contributing their labor to various activities in support of conservation action can provide a powerful incentive for public support for the cause of conserving biological diversity. Such volunteers demonstrate their willingness to pay for conserving biological resources through providing their labor—which has a calculable value—for a wide range of activities. The following are examples of what can be done:

- Volunteers, especially from nature clubs and university service groups, can help build trails and maintain facilities; appropriate supervision is required from area management.
- Professional-level volunteers can help sort out legal measures of land use, design of park facilities, etc.
- Vocational students can be given a unique opportunity to build an entire park facility themselves, requiring only that the materials and supervision be provided by the protected area managers.
- Private enterprises will often provide support to visitor centers, help design of brochures, donate materials, etc., as a public service. Their contributions can be encouraged by tax deductions, public recognition, etc.
- Businessmen's service clubs, such as Lions and Rotary, will often make donations in support of worthwhile activities.

All such use of volunteers requires that the protected area management be in a position to receive their help. The manager needs to know where to go to seek assistance; policies are required which will enable the area to receive voluntary assistance; and the activities need to be carefully coordinated with the management plan for the area (so that the tools and materials are available, plans have been drawn up, and supervision is assured).

The Special Case of Forestry Departments

The tropical forests are the home of the greatest biological diversity on the planet, supporting well over half the planet's species of plants and animals on only a little over six percent of the land area of the globe. But they are typically harvested for only a single product, timber, with the other potential products very poorly developed. IUCN has recently published detailed guidelines on the management of these lands, from which the following is drawn (Poore and Sayer, 1987).

Investments in forestry, particularly those made by forest enterprises and some international aid agencies, have often been based on simple financial analysis of the value of timber produced per dollar invested. The predominance of this approach in many tropical country forest departments has resulted in a shift of their activities away from natural forest management towards plantation forestry. Applying a broader economic analysis, where the various costs and benefits incurred by both local and distant users are included, greatly strengthens the case for investing in natural forests which are managed for a sustainable production of timber. In spite of this, natural forest management programs receive very little international financial support and are tending to decline in importance in many countries.

Natural forest management will also provide numerous other benefits to society (watershed protection and a variety of non-timber forest products). Although timber needs could often be met from plantations for similar levels of investment, the multiple benefits of the forest would not then be safeguarded against competing land uses. Where options still exist, countries should attempt to derive the maximum of their timber needs from a managed "natural forest estate."

In many countries, large-scale deforestation did not begin until the central government asserted ownership over forest lands previously held by individuals and local communities. Perhaps some lessons can be learned from this historical experience. Local people are often best able to carry out selective logging with sawing in the forest and use of animal traction to transport timber to roads and rivers; such methods have worked well in the past and should be retained as a management technique in sensitive catchment areas. As suggested in Chapter 4, when local communities are provided with the responsibility for adjacent forest lands and the products from these forests, and are able to benefit economically from sustainable levels of harvesting, then they will have a far greater interest in conservation.

Case Study 3 illustrated some of the major reasons tropical forests are being depleted. Incentives to conserve the resources contained in tropical forests include:

- charging realistic rents for production forests;
- including provisions in concession agreements for the conservation of biological diversity;
- ensuring that the income generated by the exploitation of forests and wildlife is used to provide a solid base for managing biological diversity on a sustainable basis;
- granting timber concessions on the basis of competitive bidding rather than individually negotiated agreements; and
- providing longer-term leases where these would encourage more sustainable utilization.

In order to encourage holders of timber concessions to harvest a variety of species, to harvest large trees and thereby open the forest canopy for regeneration, and to utilize each stem cut as fully as possible, governments should adjust charges to the species taken, and charge rates per tree (rather than per unit of volume). In Sarawak, for example, specific forestry charges vary considerably by species, with much lower rates on low-valued trees; as a result, Sarawak suffers only about half as much residual tree damage as either Kalimantan (Indonesian Borneo) or Sabah (Malaysia's other state on the island of Borneo) (Gillis, quoted in Repetto, 1987a).

Not all such incentives turn out to be effective. Several years ago, Costa Rica provided large tax incentives to individuals and corporations who carried out reforestation projects. It soon became apparent that most plantations thus established had been planted on good agricultural soils or were established on areas of natural forest cut and replanted to gain tax incentives. Most of the plantations were also poorly maintained.

A new program stimulating reforestation on appropriate sites by relatively small landowners who don't pay taxes has now been implemented. Investments in reforestation or natural forest management now must be on lands unsuited to more intensive uses, based on a national land use capability analysis system and also based on a management plan done by a professional forester. Further, instead of simply receiving income tax deductions for their investments, the small farmers receive tax credit certificates for the same amount. These can be sold on the open market to businesses and large land-owners who do owe taxes, at a slight discount below face value (Barborak, 1988a).

CONCLUSIONS

Policy support at the national level is essential to the success of community-level incentives for conserving biological resources. This support is required both by government policy makers and the resource management agencies who are assigned responsibility for implementing policy. Mechanisms for coordinating the various government sectors which have impacts on biological resources are also required; the national conservation strategy is an excellent means for initiating such a mechanism.

Incentives at the national level can provide much-needed support to national objectives for conserving biological resources. However, Chapter 4 showed that government action is by no means the only or even the dominant factor in conserving biological resources. The role of governments should be to establish standards for sustainable use; to design incentives and disincentives which prevent short-term gains at the expense of long-term capacity to support sustainable development; and to assist in maintaining these capacities through legislation, education, research, training, and other means.

Considerable responsibility should also rest with those agencies and institutions, especially large businesses, with the power and resources to use biological resources in sustainable ways.

CHAPTER SIX

INTERNATIONAL ASPECTS OF INCENTIVES SYSTEMS

INTRODUCTION

Biological diversity is a public good, and species and ecosystems in one part of the world can provide significant benefits to distant nations. The USA is an excellent example, as some 98 percent of its crop production is based on species which originated elsewhere. If Americans had to live on their indigenous plant species, they would have to be content with a diet of pumpkins, squashes, grapes, blueberries, wild rice, cranberries, pecans, and a few other fruits and nuts. Many of today's agricultural staples come from the tropics, including corn, rice, potatoes, sugarcane, citrus fruit, coffee, peanuts, and a wide variety of other spices, fruits, and vegetables; as the *World Conservation Strategy* pointed out, wild relatives of commercial species must continuously be crossbred with the various cultivars to improve durability, resistence to pests and diseases, crop yield, nutritional quality, and responsiveness to different soils and climates (IUCN, 1980).

All in all, it appears that far greater benefits from conserving native gene pools, especially in the wilds of the tropics, will be gained by wealthy temperate nations than the often poverty-stricken nations doing the conservation. Further, much of the depletion of biological diversity over the past 400 years or so has been caused by powerful global forces, primarily driven by markets in colonial, and then industrial, countries. Because the international community as a whole benefits from conservation, it has a distinct responsibility for sharing the effort required to conserve biological resources.

However, as Barborak (1988b) has pointed out, economic stabilization funding from bilateral and multilateral aid agencies now far exceeds development project funding in many countries facing large foreign debts, declining export prices, increasing import costs, and GNP increases which are slower than population growth. As a precondition for economic stabilization funding to pay for essential imports and to service foreign debts, developing countries are pressured by aid agencies to carry out far-reaching reforms, often involving a reduction in government payrolls and operating budgets. Agencies which have little to do with defense or essential public services are especially hard hit by such policies.

The "debt crisis" may therefore be one of the most pervasive disincentives for conserving biological resources. The reduction in ranger forces, deterioration of wildlife research programs, elimination of vital environmental education efforts, and reduction of operating budgets for conservation agencies often can cripple an entire nation's conservation efforts. On the other hand, the debt crisis also may hamper industrialization, road-building programs, and other expensive investment projects which can have negative effects on biological resources.

Since significant benefits from conserving biological resources in the tropical forests are received by the people of the industrial world (through both consumptive and non-consumptive values, including option and existence value), they should be willing to help pay for effective conservation. An important means for doing so is through the provision of economic incentives and disincentives, including direct incentives such as grants, loans, subsidies, debt swaps, and food; and indirect incentives such as commodities agreements, technical assistance, equipment, and information. Development assistance often contains a package of such incentives, including very abstract incentives such as peer pressure and public image.

It can be seen that this is a vast area, involving the highest levels of government-to-government relationships as well as more mundane people-to-people linkages. The discussion here will focus on international initiatives that are relevant to the design, implementation, and evaluation of development projects affecting biological resources.

THE ROLE OF DEVELOPMENT ASSISTANCE IN PROVIDING INCENTIVES FOR CONSERVING BIOLOGICAL RESOURCES

External capital, including development assistance funds, allows a country to invest more than it could if it had to rely on its own savings. Such funding is often utilized to cushion internal and external shocks ranging from harvest failures to major changes in commodity prices. It helps to strike an appropriate balance between reducing budget deficits and financing them, consequently diminishing certain political and social costs (at least in the short run). External capital also provides a direct cash incentive for building infrastructure, developing institutions, and transferring technology which can conserve biological resources.

Whether used for capital formation or for structural adjustments of the developing country economies, development assistance programs influence the quality of the natural and human resource base and therefore affect whether development is sustainable. It is becoming increasingly clear that many of these development programs cannot be sustained in the long term and often result in significant over-exploitation of biological resources.

Since development assistance programs of the donors are separately managed from their commercial trade, private overseas investment, and other multilateral programs, coordination is often lacking between the type of flow, its timing and its sectoral impacts. For example, the impact of food aid programs in the drought-stricken nations of Africa is often in conflict with ongoing food production activities funded by development assistance, which in turn may conflict with projects to conserve biological resources. In some countries, trade regimes promoted by a donor may encourage depletion of biological resources while exacerbating balance of payments and debt problems of the recipient country.

However, it is apparent that the citizens of most industrialized nations are in strong support of their development aid agencies providing economic incentives to assist developing countries to conserve biological resources. As this paper has suggested, numerous possibilities exist for using development assistance to provide economic incentives for activities aimed at conserving biological resources at the community and national levels in developing countries. These projects are likely to be most effective when they are part of a larger effort; single projects, no matter how well designed and implemented, are unlikely to have much of an impact by themselves.

All forms of development assistance should take into consideration their impact on biological resources. Donors should recognize development assistance as an integral component and not separate from their wider economic relations with the recipients, publicly declare their commitment to sustainable use of biological resources, and ensure that sustainable development goals are articulated in the individual donor policies when assistance is pledged at donor consortium meetings and when bilateral agreements are signed or reviewed.

Projects aimed specifically at conserving biological resources should be integrated into larger programs, and whenever possible linked with larger projects in other sectors.

In seeking to support projects to conserve biological resources, development assistance agencies should consider supporting both national and international NGOs as agents of implementation. Block grants—direct cash incentives—to international NGOs, regional organizations, or national environmental organizations could be used to further a wide variety of environmental activities at a much more reasonable cost than if undertaken as part of project activities. NGOs often have the flexibility and the local knowledge to serve as agents of conservation at the community level in ways that no government bureaucracy ever can. International NGOs can also ensure that through one relationship, an aid agency can support and work with a large number of small initiatives on the ground, without building a large bureaucracy at home.

Bilateral or multilateral development agencies, whether public or private, might consider the following types of activities in support of incentives to conserve biological resources:

- Assistance in designing community surveys aimed at discovering consumptive uses of biological resources and determining the types of incentives that might be most effective.
- Support for policy studies to determine national objectives for the conservation of biological resources; this can take the form of support for national conservation strategies, protected area system plans, and other such measures. Technical assistance in reviewing national policies can often be provided as part of larger projects, and should emanate from

the relevant economic ministries or national planning agencies. Efforts should be made to ensure that such policy reviews are based on the best available technical expertise, and are not dominated by political concerns.

- Support for assessments of biological resources through demonstrating methodologies, providing training opportunities for taxonomists and biologists, and subsidizing publication of status reports. Universities, research institutions, and NGOs need to be strengthened so that they can help governments to assess their biological resources.

- Support for information centers on biological resources at various levels—local, regional, national, international—to ensure that the information is available where it is needed, whether in a single area (such as a national park) or more widely. In particular, national databases managing information on the resources of the country should be implemented as part of a full National Conservation Strategy.

- Ensuring that all major projects include education and "outreach" elements promoting the application of economic incentives at the community level.

- Working together with other international agencies having an interest in the conservation of biological resources, including other development aid agencies, governments, the UN system, and various NGOs, to prepare global overviews on the status and management of biological resources. These overviews can be an incentive to action by these agencies, stimulating greatly increased flows of funds and other kinds of support.

- Building economic incentives measures into large development projects which affect biological resources, and promoting better cross-sectoral collaboration in such efforts.

- Providing resource management agencies with support for establishing a post with responsibilities for linking biological resources with other development sectors in productive ways, including using community development as an incentive for conserving biological resources.

- Ensuring that any projects involving protected areas include elements of economic incentives for the local communities.

- Providing technical advice on enhancing the information component of incentives packages.

- Providing financial and technical assistance to national and community development projects based on biological resources, such as development and marketing of products from native plants and animals, domestication of wild resource species, and development of wildlife tourism.

INTERNATIONAL LEGAL INSTRUMENTS AS INCENTIVES TO CONSERVE BIOLOGICAL DIVERSITY

Biological resources seem to be relatively well protected by international law, with such instruments as the World Heritage Convention, the World Charter for Nature, and conventions on wildlife trade, migratory species, wetlands, and oceans. These instruments often provide effective disincentives, in proscribing certain activities in World Heritage Sites, trade in certain endangered species, and inappropriate development in Wetlands of International Importance.

They also provide direct incentives in cash and kind, particularly through the World Heritage Fund (which includes a relatively modest budget of about $1 million per year for natural heritage projects) and through projects funded in collaboration between UNEP and the various Trust Funds established under Regional Seas conventions and protocols; Case Study 24 demonstrates that such instruments can also generate considerable national funds.

Indirect incentives are perhaps more pervasive in international legal instruments, including regular meetings of Parties (providing an incentive to have something positive to report to peers), the possibility for high-level interventions, training, and technical assistance. The Convention on International Trade in Endangered Species (CITES) both limits trade in some species and facilitates sustainable levels of trade in other species; it therefore provides important incentives for action to bring species back to productive levels and to manage productive species on a sustainable basis.

The emerging international legal instruments go far beyond simple regulatory mechanisms (disincentives) in establishing international law as a system within which States and other actors conduct their affairs rather than merely as a mediating technique between sovereign entities. This approach regards States as collaborators in a system, one of whose objectives is the sustainable management of the earth's resources.

This new perspective has immense consequences, not least in the areas of sovereignty. This is a particularly important issue, as many governments interpret their sovereignty over their biological resources in a way that hinders the recognition of the world-wide interest in biological resources, and limits the possibility of using this interest to generate financial incentives for more effective management of these resources.

Since ecosystems do not stop at national borders, transfrontier, regional and global dimensions must be reflected in adequate systems of incentives and disincentives. Indeed, the extent of the issue, rather than the source, dictates the level at which measures must be taken. From a legal point of view this simple fact has two far-reaching consequences for biological resources:

- mechanisms need to be established whereby the use of biological resources that are shared between a limited number of nations become at a minimum the subject of consultation, and at a maximum the object of common management between the States concerned (see Case Study 25); and
- mechanisms are required for managing biological resources that are of common interest to humanity, under global agreements which ensure that these resources are used sustainably, and are thus conserved in perpetuity.

Issues which should be dealt with at regional or sub-regional level are numerous, as they are determined by the distribution of the biological resources with which they deal. Agreements already exist which apply to a defined region or to States which share a common interest and have similar levels of economic development and political systems (e.g., river basin treaties). But the use of such agreements to provide economic incentives should be greatly encouraged and generalized.

Agreements of this sort can benefit from the existence of regional infrastructures (usually developed for other purposes) but onto which conservation elements can be grafted. The principal example of this development is the series of Regional Seas Agreements promoted by UNEP. These agreements are usually of the "framework" type, consisting of a number of broad general obligations to be accepted by the Parties, but leaving the details of the implementation of those obligations to be elaborated in a series of protocols or sub-agreements. This has the advantage of enabling States in the region to progress at a realistic pace dictated by what they consider they can achieve.

Both global and regional agreements on biological resources can provide useful support to development projects aimed at conserving biological resources. They can add both direct and indirect incentives, including the very powerful incentive of public opinion.

Current legal approaches to the preservation of biological diversity are piecemeal, with no binding international instrument to promote the preservation of biological diversity *per se*. A new instrument is required to define the general obligations that should be accepted by States to conserve biological resources held within their territory, the ways and means for guaranteeing appropriate access to genetic resources, and the development of mechanisms for equitable payment for use of genetic resources. The instrument might also include systems of incentives and disincentives for promoting sustainable use of biological resources, both direct and indirect.

Any new conservation convention should contain mechanisms for assessing the performance of Parties in implementing their obligations, and existing agreements which do not do so should be revised. Such mechanisms include:

- *Regular meetings of the Parties.* In addition to providing an opportunity for peer-group pressure on defaulting States, these attract a degree of visibility and media attention and permit continuous development of implementation policy guidelines through, for example, resolutions.
- *Reporting requirements.* These should be accessible to the public and linked to a meeting of the parties, at which they should be open to discussion and comment.
- *Permanent secretariat.* This provides a source of momentum specific to the convention, external to the Parties themselves, but operating at governmental level. It also provide a focus for NGO actions.
- *Significant funding.* This is the key to effectiveness of any such system of international incentives and disincentives for conserving biological resources.

Governments and development assistance agencies should also encourage more informal international legal instruments as incentives and disincentives for conserving biological diversity. These efforts might include:

- Making maximum use of protocols. If the umbrella convention contains only broad obligations, precise responsibilities (perhaps only relating to particular problems) can be more rapidly developed through the use of protocols.

- Including technical matters in an annex to the convention which can be altered by a less formal process, such as decisions of the meeting of the Parties.
- Treating conventions as binding between those States already Parties, even when insufficient ratifications have been received to enable the convention to enter into force.

CONCLUSIONS

The international community has considerable interest in conserving biological resources, both to ensure a continuing flow of goods and services for sustainable development and for more abstract existence and option values.

Two major means for the international community to provide support to incentives packages are through development assistance and international conventions. Both of these should be seen as mechanisms for expressing international support for national and community efforts to conserve biological resources.

CHAPTER SEVEN

MECHANISMS FOR FUNDING INCENTIVES PACKAGES

INTRODUCTION

Conserving biological resources requires investments. As suggested in Chapter 2, these investments are often very sound, showing high benefit-cost ratios; the more effective the economic analysis, the higher such ratios are likely to be (USAID, 1987). Chapter 4 showed how incentives packages can bring considerable sustainable benefits to local communities, and the case studies demonstrate how such systems have functioned in various settings.

However, current conservation programs are usually implemented through resource management agencies whose budgets are generally insufficient to implement their mandates effectively, and are subject to considerable fluctuation. To produce acceptable results and become truly operational, an incentives scheme must have sufficient and reliable sources of support. The following points need to be considered:

- Incentives which often come from government budgets can include national bank loans, subsidies, initial contribution to revolving funds, the government portion of shared costs, education and training, etc. Sometimes existing facilities and redistributed staff resources are sufficient, as in regularization of land tenure.
- Some incentives involve little more than an administrative decision or regulation. The cost of the action is solely the monitoring needed to ensure compliance with its terms. In

such cases the loss of tax recovery in government budgets must be taken into account. Secure land tenure comes under this heading when action consists solely of enactment of a law.

- Some incentives involve bilateral agreements or cooperation with international agencies, such as food for work programs; in many developing countries, large externally-supported development projects can often include elements which support incentives for conserving biological resources.
- Some incentives require monetary policy action prior to implementation, such as credit issued by a private bank. In the USA, for example, banks, conservation groups, and Treasury Department officials are currently discussing whether the face value rather than the discounted value of foreign debts held by US banks will be tax deductible; a deduction of the face value would give private banks a larger incentive to engage in debt-for-nature swaps (see Case Study 22).
- In some cases, community development activities are already being planned or implemented in communities which are located in or near areas important for conserving biological resources, in which case linkages with changed behavior toward conservation can be incorporated with little additional cost.
- Various non-fiscal incentives can be provided by the private sector or the general public; several of these were discussed in Chapter 5.

It is apparent that any funding mechanisms will need to emanate from the competent government authority, either in terms of enabling legislation or administrative fiat. Case Study 17 shows how one government—Costa Rica—has established a set of policies for ensuring appropriate means of financing its conservation programs.

HOW TO FUND INCENTIVES PACKAGES

While each country has its own legislation and its own ways of raising funds for conservation of biological diversity, the current period of budgetary restraint calls for innovative solutions to old problems. Case studies 4, 5, 8, 11, 17, 20, 22, and 24 contain details of funding mechanisms that have worked in Thailand, Zimbabwe, Kenya, Costa Rica, Zambia, Ecuador, and Australia. From these and other sources, the following potential sources of funding can be identified:

1. The *regular national budget* with an annual allocation consistent with the objectives and length of the incentives and disincentives proposed.
2. *Special budgets* for the initial contribution to national funds or regional revolving funds for supporting incentives packages.
3. *Charging entry fees* to national parks. This would appear to be an essential measure of the public's willingness to pay for conservation of biological resources. Funds thus earned should be returned to the protected area for management, including support for various economic incentive packages in surrounding villages (see Box 11 for further discussion of this issue).
4. *Returning profits* from exploitation of biological resources to the people living in the region. Biological resources earn profits from tourism and harvesting, so creative ways and means need to be found to ensure that a fair share of these profits are returned to the local people who are paying the opportunity cost of not harvesting the resource themselves. The case studies from Kenya, Zimbabwe, and Zambia illustrate three ways that this is being done in Africa. Protected areas should earn a fair return on the money they bring into the economy, through tourism and other means. Mechanisms may include bed taxes for tourist hotels, admission fees for national parks, departure taxes at airports, etc. Many of these are already being tapped by governments to cover other expenditures; the point is that a more equitable return needs to go to conserving the biological resources which are bringing in the funds, even when the benefits of conservation are indirect.
5. *Profits from investments* made by a protected area can often be important, where national policies permit such investments by a public agency. Janzen (1988) suggests that tropical conserved wildlands can diversify their endowment portfolios through the ownership of agricultural lands adjacent to the protected area; the agricultural profits would support management of the area. This has the ancillary benefit of the protected area controlling the kinds of agriculture carried out on adjacent lands, thereby providing a public showcase on the relationship between protected areas and agriculture.
6. *Community enterprises* based on sustainable use of biological resources, and which are part of a larger program of

conservation, can be set up to generate sufficient income to finance certain kinds of works. Apart from initial research and development expenditures (which may be recoverable), community enterprises based on biological resources—such as development and marketing of products from native plants and animals, domestication of wild resource species, and development of wildlife tourism—are essentially self-financing.

Box 11. Entry Fees for Protected Areas

Entry fees demonstrate the willingness to pay on the part of visitors; Galapagos National Park, for example, charges a fee of $40 per visitor, which is still a tiny proportion of the total price the visitor is paying for the experience. Strangely enough, many national parks do not charge entry fees, often because they do not want to discourage visitors who cannot pay and because they feel that they are providing a public service; parks are viewed as "merit goods" to which access is not denied on the basis of income. However, as costs of protected area management rise and budgets fall, most protected areas will need to consider charging fees.

In determining the fee structure to charge for the various goods and services provided by biological resources, the following points should be considered:

- What is the objective for charging fees? To supplement the regular government appropriation, or to enable the facility to be totally self-sufficient?
- How should the scale of fees compare with commercial institutions offering similar goods or services?
- How should the fee structure deal with special groups, such as children, school groups, senior citizens, low-income groups (especially local people), and foreign tourists?

The fee can be computed on the basis of actual cost of the good or service (when this can be determined); direct operating expenses, including staff; interest and amortization of investment; support for the efficient management of the area, including necessary improvements; maintenance costs; or simply what the market will bear.

---

7. *Water use charges* from irrigation projects or hydroelectric installations whose water comes from a protected area can be both justifiable and useful, improving efficiency and equity of water use as well as generating funds for protecting the watershed. This may require studies to quantify the benefits the protected area is providing; in an example quoted earlier, Hufschmidt and Srivardhana (1986) showed that annual expenditures of $1.5 million would be justified in terms of benefits to the Nam Pong reservoir in northeast Thailand. In Indonesia, the World Bank invested over $1 million to establish the Dumoga-Bone National Park to protect a major irrigation project (McNeely, 1987); water charges could be imposed to ensure that the running costs of the national park are met from the goods and services it is providing to the local community.

8. *Special taxes,* such as taxes on timber extraction, wood trading, trade in wildlife and wildlife products, concession rights or other activities connected with the sector can generate income which is then invested within the sector. This can be made more flexible by allowing tax payers to invest the amount in the kind of works which the tax is intended to promote. Special taxes can be used to set up development funds or national financing funds, e.g., for credit. An interesting example from the Ivory Coast involves creating an Environment Fund using taxes imposed on ships, especially oil tankers, docking in the country; 50 percent of the tax goes to the Fund, which is then used to purchase equipment necessary for monitoring ecosystems, preventing pollution, or improving environmental management. Since its inception in 1986, the Fund has brought in about $300,000. In developed countries, the dollar amounts involved can be far larger. For example, Florida's Recovery and Management Act establishes a Hazardous Waste Management Trust Fund to finance the correction of pollution problems should they occur. The Fund is financed by a four percent excise tax on disposal until the accrual reaches $30,000,000 and two percent thereafter.

9. *Linkages with larger development projects* can often be the best approach in developing countries. In 1985, the World Bank promulgated a major new policy regarding wildlands, with elements specifically designed to build components into large

projects—primarily for agriculture, livestock, transportation, water resources development, and industrial projects—for ensuring conservation of biological resources. These components can include economic incentives for local communities affected by the project (Goodland, 1988).

10. An *"environmental maintenance tax"* can be established as part of major development projects supported with external funding. Projects to build dams, irrigation networks, and roads might include explicit allocation of funds for thoroughly assessing the diversity of the area (thereby also supporting the development of local capacity to carry out such surveys), identifying and managing protected areas, and establishing a self-sufficient "endowment fund" for the continued management of the area.

11. A variant of such linkages is the *obligatory investment* of a percentage of the total costs in large-scale works which are dependent for their existence on environmental protection (water resources developments being the outstanding example). Sometimes an additional 10 percent allocated to reforestation and conservation works can lower the annual operating costs by increasing the useful life of the works and reducing requirements for maintenance.

12. *"Swapping debt for nature"* has proven useful in Bolivia, Costa Rica, Ecuador, and the Philippines (see Case Study 22). This mechanism involves a conservation organization (WWF, Conservation International, National Wildlife Federation, and others have been involved) buying a country's debt notes which are being discounted on the secondary market. These notes are presented to the debtor country in exchange for local currency in the amount of the face value of the debt, with the local currency being invested in conservation. While this mechanism is most useful in countries whose debts are heavily discounted (and therefore penalizes debtor countries which have sound financial management), it is still useful in a number of countries with significant biological resources.

13. *Building conditionality into extractive concession agreements* can be an effective instrument in countries which have such extensive timber or fisheries resources that concessions are sold to private investors. As part of such agreements, the

concession holder could be required to provide support to various incentive programs aimed at maintaining the long-term productivity of the area being logged or fished. Where concessions are given for forest use, governments must ensure that they realize a significant proportion of forest rents and that, as a minimum, a proportion of such rent is returned to managing the forest to ensure its long-term productivity. In general, governments should design incentive systems which encourage sustainable use of the biological resources of the forest ecosystems.

14. *Profits earned from non-extractive concessions* can often provide sufficient funds for running a protected area, as from hotels, tours, and restaurants. Such concessions should be granted on the basis of conditions that do not detract from the natural values of the protected area, and the profits from such concessions should be returned to the resource management agency. Such concessions might also be required from tour companies bringing tourists into protected areas, even if they do not stay overnight; this could supplement admission fees.

15. *Encouraging voluntary support from the private sector,* especially those involved in resource extraction or in non-consumptive uses of biological resources (such as tourism) can be effective, though such voluntary support is difficult to predict and incorporate in planning efforts. Such voluntary support might be particularly appropriate where a number of tourist enterprises rely on protected areas for their livelihood (see Box 12 for an example from Nepal).

16. *Direct support from development assistance agencies* is often feasible when the living conditions of rural people are to be improved (recalling that many of these are the "poorest of the poor" and therefore of particular concern to many bilateral government agencies, and to various church, population, and food related PVOs). A major point here is that effective incentives packages seldom require major funding, but rather effective funding aimed at very specific targets; therefore, development assistance agencies may need to aggregate a significant number of community-level projects in order to attain the project magnitude that is administratively attractive. The major drawback to this approach, is that it may

breed dependence rather than self-reliance, unless the support is provided with great sensitivity.

Box 12. Private Sector Support for Conservation in Nepal

In many countries, the private sector provides significant incentives for conservation by providing grants to activities which lead to enhanced management of biological resources. One outstanding example is the International Trust for Nature Conservation, established by the Tiger Mountain Group (a nature tourism organization operating primarily in Nepal). This trust was designed to recycle excess profits from nature tourism into activities which would promote the protection of wildlife and its habitat.

One of the principal activities has been a conservation education program aimed at the villages that surround Royal Chitwan National Park, where Tiger Tops Hotel is the flagship of the Tiger Mountain Group. More recently, the scope of the Trust has been expanded to include more general concern with sustainable development in the areas surrounding the Group's operations. The Trust is putting into practice its belief that wildlife must increasingly pay for itself if it is to survive in today's crowded world (Roberts and Johnson, 1985).

---

17. *Direct support from international conservation organizations* has tended to focus on the biological resource rather than the people, but this is beginning to change and organizations such as WWF, Conservation International, New York Zoological Society, Frankfurt Zoological Society, The Nature Conservancy, and many others are now becoming more aware of the linkages between people and conservation. Such organizations can often provide at least seed funding to get appropriate incentives projects started, and they have been involved in a number of the case studies. IUCN, through its work in national conservation strategies, may be able to promote funding mechanisms being developed for incentives packages. Finally, private conservation agencies may have access to blocked funds owed to private companies operating in developed countries, and be able to apply such funds to incentives packages.

18. *Local currency counterpart funds* derived from PL 480 (a US Public Law which enables certain nations to pay in local currency for food imports from the US, with the local currency to be spent in the importing nation) and other public-sector international assistance operations can often be used to support conservation efforts, including incentives packages. Kux (1986) has pointed out that for USAID, at least, it should be relatively painless to increase investments in conservation considerably through greater use of local currencies generated from sales of agricultural commodities provided by the USA to some developing countries. These funds could be used for activities such as the purchase of land for protected areas, for inventories of tropical forests, education and training, and support for alternatives to destructive land use practices.

19. *Donations from multinational corporations* investing in resource-based activities in developing countries can contribute to conservation incentives packages, both to protect their own investments and to contribute to host country conservation goals. Such donations are often facilitated if the government conservation agency, or a private institution, has established a mechanism for receiving such donations; experience has shown that private industry is less eager to provide voluntary funds to regular government programs than to an independent foundation (especially if the donations are tax-deductible).

20. In some cases, *foundations established by or for a protected area or protected area system* can be a useful stimulus for generating non-governmental sources of funding (many of which might come from sources discussed above). In Indonesia, for example, the Indonesian Wildlife Fund is supported by voluntary contributions from the timber trade; it was established by the Ministry of Forestry but operates independently under a board of directors who allocate the funds in support of various conservation projects. In Zambia, an essential element in the success of its Wildlife Fund has been its establishment within the National Parks and Wildlife Department (Case Study 20). Additional elements to consider in establishing such a foundation or trust fund are covered in Box 13.

Box 13. Establishing A Mechanism to Receive Donations

Donations from the public or private enterprise in support of conserving biological resources often depend depend on there being an acceptable organization to receive such donations, and this may require a foundation, trust fund, or endowment fund.

Foundations must have a legal basis, perhaps as a corporation or limited liability charity, and government policy must be designed and implemented to enable the establishment of such a legal entity by or for a government agency. Further, donations to the Foundation must be made tax-deductible.

A Foundation (or Trust) for either a specific protected area or an entire system for conserving biological resources depends on community approval for the proper operation of the protected area(s) for which outside financial support is being sought.

The Foundation must have a clearly-stated objective (such as "To develop and improve public facilities for conserving biological diversity"). It might be willing to accept contributions either to its general fund, or for contributions aimed directly at certain parts of the management plan. In providing funds for conserving biological resources, the Foundation has three major options: grants for specific activities; subsidies for running costs; and loans for facilities which are self-liquidating (where the principal and interest are returned to the Foundation from the net profit from the operation).

Major contributions to such a Foundation or Trust should be based on two basic conditions:

- Contributions must go to the Foundation or Trust rather than directly to a government agency. The Foundation must be governed by a Board which is above reproach, and audited accounts must be publicly available.
- Contributors must be able to actually see the result of their contribution, so projects must be packaged creatively to have qualities of excellence, significance, and uniqueness. Projects which benefit children, are of broad significance to the community, and have public appeal are far easier to "sell" than are the more mundane (but no less important) matters such as routine maintenance and running costs.

21. *"Conservation concessions,"* parallel to those for forestry or mining, might be provided to international conservation organizations for areas of outstanding international importance, in exchange for a rent which would be provided to the resource management agency for funding other areas. The concession agreement would specify standards of management, access to the public, permissable developments (usually non-extractive), etc., and the international agency would assume full responsibility for living up to the concession agreement. Development aid agencies might consider providing support to local NGOs or other agencies for purchasing concessions on a few outstanding areas, and developing them as a demonstration of how an area can be developed so that its biological resources can be managed in an economically sustainable manner.

22. As a variant, *property rights for species or protected areas* of outstanding importance might be issued to conservation organizations or relevant UN agencies, with payments being made to the government and the concession holder being required to manage the species or area to a high international standard (and subject to a contractual agreement with the government).

CONCLUSIONS

In general, incentives packages should be supported to the maximum extent possible through the marketplace, but the marketplace needs to be established through appropriate policies from the central government.

The problem faced by all of these funding mechanisms is that they face opportunity costs; any funds earned might be used by the government in other ways that the government considers of higher priority. The attraction of the methods suggested in this Chapter is that the income is being earned by the biological resources, and much of the funding is being provided by the public in expression of their support for non-consumptive uses of biological resources.

The major requirement from government policy makers is that they recognize the many values of biological resources, and take advantage of opportunities to invest in the continued productivity that such resources require.

CHAPTER EIGHT

GUIDELINES FOR USING INCENTIVES TO CONSERVE BIOLOGICAL DIVERSITY

INTRODUCTION

The preceding discussion has described how a system of incentives can work to conserve biological resources at community, national, and international levels, and suggested sources of funds for such systems. Particular attention has been given to the policy changes which are required at the national level and the support required from the international level in order to enable incentives to work at the community level, and to describing the incentives that are available to bring about such policy changes.

The major constraints faced by an incentives system include:
- the long period of time between investment in conservation and return on the investment;
- short-term hardships caused to subsistence resource users who lack alternative livelihoods;
- lack of information on the economic benefits of conservation;
- lack of sufficient financial resources for conservation, especially in developing countries;
- the problem of benefits from conservation accruing to other countries (international externalities);
- low political payoffs from investments in conservation; and
- weakness of government institutions at local level, with resulting inability to implement effective management.

The problems are so serious that governments must take decisive action, and accept that some additional investments will be required; but sustainable development of biological resources will likely be cheaper than rehabilitation programs, and many—even most—conservation efforts have proven cost-effective on the basis of traditional economic criteria.

Action is required at two general levels: The regional or national plan; and the specific project. The first is strategic, establishing national objectives for addressing on a broad front the fundamental problems of degradation of biological resources. The second is more tactical, attacking specific parts of the problem with action tailored to the needs of the situation. The procedure for developing and implementing incentives is quite different in the two cases, but each is dependent on the other for its success.

The following guidelines are intended to stimulate the greatest possible government commitment to conserving the entire spectrum of biological diversity, in an economically optimal way; and to assist development agencies—both national and international—in improving the design of projects that affect biological diversity. They provide practical advice for the formulation of policies for the sustainable development of biological resources, and for the conversion of policy into practice through specific project interventions. They include detailed advice on how incentives packages can be designed and implemented by resource management agencies, and how specific project interventions can be most effective.

GUIDELINES FOR CENTRAL GOVERNMENT PLANNERS

INTRODUCTION: *Why Incentives Are Required to Conserve Biological Resources*

Some of our planet's greatest wealth is contained in the species of plants and animals living in natural forests, plains, mountains, wetlands and marine habitats. While this wealth has great potential for supporting sustainable development, problems have arisen as governments and local populations have increased their demands on the biological resources. Since future consumption of goods useful to humanity depends to a considerable extent on the stock of natural capital, conservation is a precondition for sustainable development.

But instead of conserving the rich resources of forest, wetland, and sea, current processes of development are depleting many biological resources at such a rate that they are rendered essentially non-renewable.

Many of these resources have considerable market value, and if managed appropriately their sustained productivity càn help support rural and urban communities far into the future. Effective systems of management can ensure that biological resources not only survive, but in fact increase while they are being used, thus providing the foundation for sustainable development and for stable national economies. Significant political benefits can ensue.

The fundamental constraint is that some people earn greater immediate benefits from exploiting biological resources than they do from conserving them; society at large often pays the costs of such resource depletion. To the extent that resource exploitation is governed by the perceived self-interest of various individuals or groups, behavior affecting maintenance of biological diversity can best be changed by providing new approaches to conservation which alter people's perceptions of what behavior is in their self-interest. Since self-interest today is defined primarily in economic terms, conservation needs to be promoted through the means of economic incentives.

It is apparent that conserving biological resources requires appropriate government policies in many sectors, and that using economic incentives will not bring about miraculous cures to society's conservation ills. However, economic approaches can help clarify issues and indicate costs and benefits of alternative courses of action, providing an important tool to governments that are concerned about managing biological resources more effectively.

Since governments establish the policy framework within which individuals and institutions operate, they should ensure that the resource management agencies have the policy support which will enable them to carry out their assigned responsibilities. Since human decision-making is usually based on economic thinking, the benefits of linking economics more explicitly with the conservation of biological resources are manifest.

GUIDELINE 1: **MAKE RAPID INITIAL ASSESSMENT OF AVAILABLE BIOLOGICAL RESOURCES**

In order to develop informed policies on depletion rates, rates of sustainable yield, national accounting systems, and land use planning, all governments should build the capacity to assess the status, trends, and utility of their biological resources. This capacity should include:

- national compilations of the flora and fauna (at least higher plants and vertebrates) contained within the nation, in addition to the more usual assessment of stocks of timber, fish, and minerals; where these compilations do not yet exist, development projects might require that rapid appraisal methods be employed—perhaps through the use of indicator species which can provide the optimal return on investment of field time—to ensure that biological resources are being given an appropriate level of priority;
- institutionalized biological surveys, perhaps carried out by university departments of biology, to determine what species occur where and in what numbers, and how these parameters change over time;
- a national program for monitoring the status and trends of biological resources, linked to international systems such as UNEP's Global Environmental Monitoring System and the World Conservation Monitoring Centre (operated by IUCN, in collaboration with UNEP and WWF); and
- regular publication of the available information on status and trends of biological resources, and the various forces which are affecting these trends.

These efforts will help governments to recognize the consequences of their development activities on the biological resources of the nation, and help identify external effects of development projects on biological resources. However, in-depth assessments are time-consuming, and action should not be delayed until "all the information" is available; instead, some rapid initial assessments need to be done. Development assistance agencies may be willing to assist in such efforts.

GUIDELINE 2: **ESTIMATE THE CONTRIBUTION OF BIOLOGICAL RESOURCES TO THE NATIONAL ECONOMY**

As a basis for applying economic incentives and calculating marginal opportunity costs, governments need to estimate the economic contribution that biological resources make to the national economy. This requires:

- ensuring that national accounting systems make explicit the tradeoffs and value judgements regarding impacts on biological resources that may not be measured in monetary terms;
- conducting research on methodologies for assessing the cross-sectoral impacts—positive and negative—of resource utilization;
- collecting information on the physical properties of resources in specific environments and for specific uses;
- developing methodologies for assigning values to non-marketed biological resources, appropriate to the needs of the country; and
- estimating the economic productivity of various ecosystems, with various types of inputs.

Governments should consider using systematically the concept of Marginal Opportunity Cost in their development planning, as a means of assessing the true costs of allowing the depletion of biological resources to continue and seeking alternative paths toward sustainable development.

The sustainable levels of production of economic benefits from biological resources, including fish, timber, wildlife, medicinal plants, and other goods and services, should be estimated and demands upon benefits planned within those limits. This should be reflected in the prices of forest products and other biological resources.

The review and formulation of all national policies which have a direct or indirect bearing upon biological resources must therefore:

- estimate the relevant benefits which biological resources can produce;
- treat biological resources as capital resources and invest accordingly in preventing their depletion;
- ensure that the objectives of sustainable utilization are met; and
- address the basic needs of the local people who depend on biological resources for their continued prosperity.

GUIDELINE 3: **ESTABLISH NATIONAL POLICIES FOR MANAGING BIOLOGICAL RESOURCES**

The incentives which are required to conserve biological resources at the community level usually require commensurate policies at the national level. A national or regional conservation strategy can be an effective means of reviewing such policies, and determining what shifts are required to achieve national objectives for conserving biological resources. Major policy components of the required integrated action might include the following considerations:

- The many economic and financial benefits of integrated rural development linked with conservation of biological resources need to be quantified and brought to the attention of policy makers.
- Both conflicts and potential for cooperation between the various activities of agriculture, fisheries, forestry, conservation and rehabilitation need to be identified in integrated plans and programs.
- Institutional reform and improvement is often a prerequisite to good design and implementation of integrated sectoral development plans and programs.
- Legislation consonant with the socio-economic patterns of the target group and the natural resource needs to be formulated, both to institute disincentives and to ensure that incentives carry the power of law.
- Policies and legislation in other sectors need to be reviewed for possible application to conservation of biological resources and community involvement in such work.
- Effective incentives need to be devised to accelerate integrated development to close any gap between what the individual sees as an investment benefit and what the government considers to be in the national interest.
- The rural population needs to be involved in the design and follow-up of plans and projects, not simply their implementation.

Systems of incentives can be designed in a large number of ways, and numerous options exist for coordinating these incentives with other national policy objectives. In designing systems of incentives, governments should compare several options, with estimated costs and benefits, for each of the various national objectives being addressed. Systems of incentives need to be supported by suitable machinery for implementing the system, including regulation, enforcement, monitoring, and feedback.

All government sectors which depend on biological resources should design policies to encourage the sustainable use of these resources, possibly as part of the process of preparing a national conservation strategy. In addition, other sectors which have major impacts on biodiversity, such as transport, highways, and the military, should ensure that their policies do not unnecessarily deplete biological diversity.

Coordination and control of natural resource use in order to handle external effects, in particular to introduce systems of incentives which involve several sectors, may require the creation of new agencies with wide-ranging authority over certain aspects of the operations of implementing ministries within a particular region.

Based on the information collected and managed following Guidelines 1 and 2, governments should establish national objectives about what are the desired levels of biological diversity. However, under current constraints of finance and manpower, it will often be necessary for guidelines 1 and 2 to be carried out simultaneously with Guideline 3, or even to follow Guideline 3. Drawing on the latest advances in genetics, population dynamics, and conservation biology, governments need to state, as a matter of public record, what proportion of the current land and water area is intended to be legally protected for conserving biological resources. Such policy objectives can often be incorporated as part of a national protected area system plan or a national conservation strategy; on the basis of such national objectives, governments can measure the costs and benefits of implementing conservation programs effectively.

GUIDELINE 4: **REMOVE OR REDUCE PERVERSE INCENTIVES**

A major step in moving from exploitation to sustainable use is for governments to analyze the impacts of all relevant policies on the status and trends of biological resources. Such an analysis would involve detailed determination of Marginal Opportunity Costs, including costs and benefits of direct and indirect values.

Based on the policy analysis, governments should eliminate or at least reduce policy distortions such as subsidies that favor environmentally unsound practices, and at the same time discriminate against the rural poor, reduce economic efficiency, and waste budgetary resources.

An analysis should be made of incentives provided to promote activities which affect lands important for conserving biological resources, including such measures as tax concessions, credit, grants or indirect incentives such as provision of infrastructure. Future incentives should be designed to ensure a more optimal, sustainable production of a range of benefits as well as an equitable distribution of such benefits.

GUIDELINE 5: **ESTABLISH A STRUCTURE OF RESPONSIBILITY FOR THE BIOLOGICAL RESOURCES IN THE REGION**

While those resources contained within strictly protected areas are usually a government monopoly, biological resources in buffer zones, game reserves, national forests, and communal properties are often "open access goods" and need to be brought into some form of resource-use control. Granting usage rights can often be an effective incentive to control the use of a biological resource of considerable national importance. Such products as firewood, medicinal plants, and meat can often be made available to local communities more effectively through direct harvesting than through middle-men, and usage rights can often provide economically disadvantaged communities with highly valued resources.

Incentives can be used to create an institutional setting in which the property rights to specific populations of species of plants or animals are held by a single decision-making unit. Communities, lineages within a community, or other forms of informal cooperatives, have often provided the basis for community-based resource management systems. Such systems have proven their relevance over time, but are now being overwhelmed by modern incentives for exploitation. To counteract this trend, governments should consider ways and means of implementing incentives which would enable these systems to become effective once again. In addition, community-based resource management systems which are functioning well—such as protection of sacred forests, water-use cooperatives, and equitable sharing of access to fisheries and grazing lands—should be strengthened through being incorporated into the incentives package.

The intention of all packages of incentives and disincentives aimed at the local community should be to ensure that the local people steadily enhance their capacity to utilize biological resources in an optimal and sustainable manner. This will often involve self-reliance built on sustainable uses of the biological resources available in the local ecosystems, and will effectively reduce the dependence of rural communities on external inputs.

GUIDELINES FOR RESOURCE MANAGEMENT AGENCIES

INTRODUCTION

Most natural resource management agencies—such as departments of national parks, forestry, and fisheries—have tended to be more concerned with the resource than with the people who are affected by how the resource is managed. Fortunately, this perspective is beginning to change and earlier chapters have demonstrated the many benefits of working with local communities, and the costs of not doing so. In most cases, local incentives packages will need to be administered by the resource management agency, or at least with the involvement of the agency.

The following guidelines suggest ways and means for resource management agencies to enhance their capacity to design and implement incentives packages, based on the assumption that they receive the necessary policy support from central government. Where this capacity requires improvement, assistance might be sought from various international agencies. Many of the guidelines for designing and implementing development projects affecting biological resources will also be relevant to the resource management agencies.

GUIDELINE 1: **DEVELOP THE INSTITUTIONAL CAPACITY FOR IMPLEMENTING ECONOMIC INCENTIVES TO CONSERVE BIOLOGICAL DIVERSITY**

Any incentives scheme must be designed within the capabilities of the relevant institutions. In seeking to develop that capacity, the management agency should ask the following questions about itself:

1. Does the agency have real coverage of the target areas, and enough staff to both promote the plan and provide the technical assistance, education, and training to carry it out? If not, can the agency gain access to the necessary staff in other ways?

2. Is the necessary inter-agency, bilateral or international cooperation within the capabilities of the executing agency staff?

3. Is the balance appropriate within the agency between headquarters managerial staff and the field staff who are actually implementing the incentives package?

4. Are the field staff sufficiently well trained to be effective workers in community development, as well as in conservation of biological resources?

5. Are the local administrative and decision-making procedures of the agency implementing the incentives package sufficiently decentralized to be effective?

6. Does the institution have solid technical and research data to support field staff?

7. Does the institution measure success by the quality of its work instead of just by meeting quantity targets?

8. Does the institution have simple, non-bureaucratic procedures with minimum red tape so that incentives can become real tools for sustainable development of biological resources?

The answers to these questions will provide the resource management agency with guidance on how it needs to develop further its capacity to implement incentives packages. The first step in this process may be to establish a "Community Development Liaison Officer," with the mandate to become familiar with the activities of all government and NGO agencies in the region and to seek ways and means of linking those activities with local and national conservation objectives.

GUIDELINE 2: **ENSURE COMMUNITY INVOLVEMENT IN THE INCENTIVES PACKAGE**

Earlier chapters stressed the point that the foundation of any incentives package is community support, and such support is gained only through involvement. The following elements are essential:

Motivation. Potential participants must be convinced that the problem being addressed by the incentives package is a high priority for the community. If farmers are shown that the proposed project can help overcome present constraints, the results will be positive. This is done by making the community part of the project planning process from the earliest stages, and making them the leading actors throughout the program.

Benefits. Both the individual farmer and the larger community must clearly perceive the benefits they will derive from the planned conservation action, either through direct profits from the action or else from the incentives themselves.

Information. The community needs to be informed about the incentives package, including its costs and benefits, and any accompanying disincentives. The implementing agencies need to clear up any doubts and encourage the rural people to participate fully. The outcome of the promotion campaign should be a better informed rural population which participates actively in conservation activities.

Viable options. The options offered to rural people need to be accessible, and within the capacity of government or private enterprise to provide. Solid financial and logistical backing must be guaranteed and any restrictions to local participation eliminated.

Skills. The rural people need to have or obtain the skills required to implement the activities stimulated by the incentives package, which implies technical assistance and training as well as education in the broad sense.

Determining which incentives will be most useful in stimulating the desired behavior at the community level, should begin with analysis of how current government social and economic policies are affecting the behavior of the villagers toward biological resources. It is often useful to undertake a socio-economic survey of the communities affected by regulations controlling use of biological resources. Such surveys can also provide the necessary raw material for determining the types of incentives that are required to bring about the desired changes in behavior. Information collected might include:

- the ethnic diversity of the communities and their social structure;
- the traditional location and proximity of householder and kin groups for ritual, labor exchange and other important community activities;
- standard indicators of socio-economic well-being, including demographic parameters such as population and age structure as well as indicators of health and education;
- the pattern of economic activity, in both time and space, particularly in regard to how this affects biological resources;
- patterns of land tenur, land use, and access to resources;
- the biological resources now being used, how the resources are being harvested, the degree of awareness about controlling regulations, and possible alternative sources of income; and
- the importance of the biological resources, both economically (food, raw materials, income) and socially (role in kin and other community relationships).

This information can provide managers of biological resources with the necessary insights into the needs and desires of the local people, and can avoid misunderstandings and disruptions when implementing incentives packages.

Such surveys can also provide the necessary information for determining the appropriate level of incentives that will move individuals to respond in the socially desirable way. They can also indicate the best means of providing incentives, ensuring that they are perceived as fair, equitable, and fairly earned. Community-level institutions should be fully involved in the design, implementation, and interpretation of such surveys.

GUIDELINE 3: **DESIGN REALISTIC INCENTIVES PACKAGES, AND MONITOR THEIR APPLICATION TO ENSURE THAT MODIFICATIONS ARE MADE IF NECESSARY**

Elements to bear in mind when designing and implementing incentive packages that are effective include the following:

1. The incentives should serve to catalyze initiative. They should be considered fair compensation for work done, and not as a gift.
2. The incentives must tend to emphasize the implementation of mechanisms and methodologies over simply supplying money in cash. Where cash is supplied, the tendency should be to invest more money in community development works.
3. The incentives package should be reviewed when new circumstances arise. The technology being used needs periodic review as well.
4. The incentives should be part of an integrated approach targeted at eliminating the battery of constraints to conservation due to local physical and social circumstances; they should help correct market failures.
5. Incentives which imply distribution of surpluses among contracting parties—such as the case of harvest of cane or meat from national parks—must be carefully and clearly regulated. No group should feel that its interests are being neglected.
6. The incentives package should produce both short-term and long-term results, the former to make them attractive to the target audience and the latter to ensure their longevity.
7. Incentives should be granted on a flexible basis. Demands with which the community is unable to comply should be eliminated beforehand.

GUIDELINE 4: **INCORPORATE ECONOMIC INCENTIVES INTO THE PLANNING PROCESS FOR THE AGENCY**

National protected area policies should include an economic justification for conserving the areas, provision for comprehensive planning and management to ensure the sustained profitability of the resource, and linkages between protected areas and other relevant sectors (such as agriculture, tourism, communications, community development, forestry, and water resources development). The management authority should specify what each protected area will provide to the national economy in terms of employment, construction costs, cost of food for picnics, fishing and camping equipment, transportation, watershed protection, and genetic resources.

In order for protected area authorities to benefit from the incentives potentially available from these other sectors, coordinating mechanisms should be established. A senior staff person might be appointed, with terms of reference for determining what opportunities exist for productive collaboration with other sectors, and particularly with community development initiatives (both governmental and non-governmental).

The development of each protected area should be guided by a long-term (five years is a useful planning horizon) management plan which specifies the objectives for the area, the management steps required for achieving the objectives, and the means currently available for implementing management, and the additional means required to implement the plan. The latter should include potential economic incentives and disincentives, and the policies required to convert their potential into reality.

Each plan should also include mechanisms for providing incentives and disincentives to local people. This section should be prepared with the full involvement of the affected communities, and should include objectives for the incentives, specify what is expected from the community in return for the incentives, and outline options for implementing the incentives.

Protected area managers should ensure that all educational and interpretive materials used in and around the area also include appropriate mention of economic relationships with surrounding communities.

GUIDELINE 5: **DEVELOP INNOVATIVE FUNDING OR OTHER MECHANISMS THAT WILL ENABLE THE PUBLIC TO SUPPORT CONSERVATION OF BIOLOGICAL RESOURCES**

Since few government conservation agencies have sufficient funding to carry out their mandates effectively, innovative funding mechanisms need to be sought outside the traditional government sector. Some of these may require policy support from the central government or ministries of finance, such as tax deductability for donations of cash, land, or services. Other options which might be considered include: charging entry fees; returning profits from exploiting biological resources to the people living in the region; implementing water use charges for the water produced by a protected area; establishing linkages with major development projects; building conditionality into extractive concession agreements; seeking support from international conservation organizations; and considering "conservation concessions," similar to those for forestry or mining.

Protected area management authorities, or those seeking to help support them, should consider the establishment of a Foundation or Trust which will support conservation of biological resources, either directly through the protected area authority or more broadly to cover all aspects of biological resource conservation.

Labor and other donations in kind can often be very useful means of enabling the public to express the value they place on the existence of certain biological resources. Protected area authorities should therefore give careful consideration to the ways and means available for encouraging voluntary community service labor for conserving biological resources.

GUIDELINE 6: **ENSURE THAT INCENTIVES ARE PERCEIVED AS SUCH**

Incentives and disincentives aimed at changing the behavior of individuals must clearly and explicitly indicate the linkage between rewards and behavior. This will usually require that effective information programs are provided to those receiving the benefits. When individuals or communities first receive an incentive, they should be informed in detail of how the incentive works and why they are receiving it. They should then be reminded on a regular basis that the benefits are flowing to them because they are contributing to national objectives for biological diversity, or live in or near an area which is of national importance for sustainable use of biological resources.

It is often useful to prepare educational material on the benefits being provided to villages around protected areas or other areas of national importance for conserving biological resources; while such material is of particular use in the schools in the villages most directly concerned, it can also be used more widely to demonstrate government commitment to conserving biological resources.

More generally, public information programs should stress the importance of the entire population helping to conserve the environmental resources that local people "harvest," including clean and plentiful water, clean air, biological diversity, and attractive scenery.

GUIDELINE 7: **INCORPORATE DISINCENTIVES AS PART OF THE PACKAGE**

While the marketplace is usually a more powerful determinant of human behavior than regulations, experience has shown that clear regulations which are understood and supported by the local community, with penalties set at the appropriate level (that is, exceeding the benefits derived from the illegal activity), are often a necessary part of the package of incentives and disincentives for local communities. Appropriate disincentives exist in most countries, in the form of laws and regulations, supported by fines and jail sentences; but national legislation is seldom sufficiently well enforced to provide a particularly powerful disincentive. When supported with appropriate incentives and by public opinion, the local community can often be an effective enforcer of disincentives. Governments need to enact policies which enable the local communities to play this positive role in enforcing disincentives.

GUIDELINES FOR DESIGNING AND IMPLEMENTING DEVELOPMENT PROJECTS

INTRODUCTION

In most countries, both governments and the private sector are already using incentives, but these incentives are not being used to support conservation. In order to demonstrate how incentives can be applied to change behavior that leads to sustainable use of biological resources, demonstration projects can be designed to address urgent problems. Demonstration projects test a full range of methodologies, and develop experience in implementation of construction works, community development, application of incentives, training, and technical assistance. Successful projects may become showcases, convincing rural people, governments, academia, and the private sector that conservation is both necessary and beneficial; they can lead to a series of replications throughout the country.

It is apparent that virtually all projects which have a component which deals with biological resources will benefit from incorporating economic incentives and disincentives into the project. The following guidelines are aimed at assisting those responsible for designing and implementing development projects, either at national or international level and with governmental or nongovernmental agencies, to ensure that all relevant matters have been taken into consideration.

GUIDELINE 1: DESIGN THE INCENTIVES AS A PACKAGE

Incentives and disincentives can seldom stand alone; they need to be part of an overall strategy or plan which includes a variety of incentives and disincentives. In selecting the elements for inclusion in such a package, the following points are pertinent:

1. Consider the factors which are universally relevant and provide the foundation for almost any kind of incentives package. These include: secure land tenure; development and strengthening of local institutions; training and education; and technical assistance.
2. Based on information gained from surveys of the target communities, design the specific package of incentives to meet the

highest priority needs of the villagers, with explicit objectives to be attained. When incentives are designed to enhance the management of a protected area, they should be closely linked to the management plan for the area. This requires that the protected area manager is fully involved in the design and implementation of the incentives package.

3. Assess the resources, including the biological resources and the human resources available for implementing the incentive. The biological resources may need to be surveyed, using local universities, research center staff, and other expertise that may be available.

4. Assess human motivation for both conservation and exploitation. What are the factors underlying current over-exploitation of biological resources, and what motivating factors are available for changing those factors? The needs and aspirations of the local people need to be discovered before any reasonable system of incentives can be designed.

5. Assess all development plans which might influence the incentives. What are the other development projects which are affecting the project area?

6. Conduct a preliminary economic analysis. What is the opportunity cost for working in a particular area or region, and how does this area relate to other areas having the same biogeographic characteristics?

7. Select the types of incentives. Incentives usually must be site-specific, but certain aspects of the incentives issue can be underscored as part of land use planning policies and plans for conservation:

    • Incentives need to be classified as general in nature or as targeted at specific priority regions in the country.
    • Incentives which are nation-wide or region-wide (such as taxes or use rights) in scope must be established and regulated by a legal body, to guarantee users that they are entitled to insist on State compliance where they themselves have complied with the established regulations.
    • Incentives in a national scheme must integrate land tenure and its regularization in such a way that the cultivators are guaranteed that they will reap the fruits of their labors.
    • Incentives must be designed to ensure continuity of plan activities even after the incentive is no longer applied.

- Incentives must be well planned and realistic. Funds must be available to back them, and they must complement one another and be carefully promoted beforehand.

GUIDELINE 2: DETERMINE THE CAPACITY OF THE LOCAL COMMUNITY TO BENEFIT FROM INCENTIVES

The capacity of any given village or community to benefit from incentives will vary considerably from community to community. The effectiveness of a package of incentives aimed at a specific community depends on a number of factors, including:

1. the major *objectives* of the incentives scheme (the most important issue here is to be very clear and explicit about what conservation objectives are to be achieved by the incentive);
2. the *capacity* of the community to absorb incentives (villages with well-developed institutions will usually be able to absorb incentives more effectively than poorly organized villages, which may first require the development of appropriate institutions);
3. the initial *state of the biological resources* to be managed (incentives to manage existing resources are different from incentives to rehabilitate resources that have been depleted);
4. the *level of motivation* of the community (communities which are eager to cooperate and take advantage of opportunities such as tourism are quite different from communities which need to be convinced that cooperation is in their own best interest; in the latter case, an initial promotion campaign may be required);
5. the *constraints* which the incentives are intended to overcome (these can include: lack of title to land; unclear responsibility for biological resources to be conserved; insufficient information about available options or rights under the law; lack of access to resources, expertise, or appropriate markets; and insufficient awareness of the benefits available from conservation action);
6. the *effect of time* on the incentives (including the time required to apply the incentive, the time over which the incentive needs to be applied, the time required for the incentive to

bring about the desired change in behavior, and the time to recover any recoverable investments); and

7. the *method of distributing* the incentive to the community (communities with strong institutions may use them to distribute the incentives, while other mechanisms may be required in other cases; this will obviously vary with the objectives and degree of motivation).

GUIDELINE 3: **ENSURE THAT PROJECTS WHICH INCORPORATE INCENTIVES INCLUDE ALL NECESSARY ELEMENTS FOR THEIR SUCCESS**

When designing or assessing a project which incorporates economic incentives, the following questions need to be answered. Any negative answers should require additional explanation; some projects will be designed to seek answers to these questions.

1. Has the project established *what are the biological resources* for which management needs to be enhanced?
2. Has the project estimated *the economic values* of the resources for which management is to be enhanced through the incentives?
3. Have clear and explicit *conservation objectives* been established for the package of incentives and disincentives?
4. Has the project identified *perverse incentives* (i.e., the national social and economic policies that have encouraged the community to over-exploit biological resources) and identified the means to overcome these perverse incentives?
5. Has the project presented sufficient *information about the community,* including determining what biological resources the community is currently using, how the resources are being managed by the community, the degree of awareness about controlling regulations, and possible alternative sources of income?
6. Does the project contain *specific packages of incentives* which are aimed at effectively meeting the highest priority needs of the villagers, and ensuring that the incentives package is linked with other development activities?
7. Does the project establish a *structure of responsibility* for the biological resources in the area? Does it build on existing village institutions, or build new ones?

8. Does the project incorporate *packages of disincentives,* through legislation, regulation, taxation, peer pressure, and appropriate levels of penalties?

9. Does the project provide appropriate *information and public education* to the target audiences on both incentives and disincentives?

10. Does the project contain a means of *monitoring and feed-back,* so that necessary changes can be instituted as the incentives package adapts to changes?

11. Will the project lead to *permanent or sustainable funding mechanisms* which will enable the incentives to continue operating after the life of the project?

CHAPTER NINE

CASE STUDIES

THE USE OF INCENTIVES AT THE NATIONAL AND INTERNATIONAL LEVELS

CASE STUDY 1: **INCENTIVES WHICH DEPLETE BIOLOGICAL RESOURCES IN BRAZIL**

A recent study by the World Bank has shown that the government of Brazil has enacted a series of policies, tax incentives, and legal rules in order to accelerate the pace of settlement in the Amazon basin, thereby leading to deforestation which is causing severe environmental problems. Other provisions encourage the conversion of forests to pasture and cropland in order to reduce the tax liability, thereby leading to excessive deforestation of marginal land on large farms. These incentives include:

Tax laws. Brazil's income tax laws virtually exempt agriculture and convert it into a tax shelter, so urban investors and corporations are competing aggressively for land to establish livestock ranches. By initiating such tax incentives and thus making it attractive for wealthy individuals to buy land from small farmers in areas of well-established settlement (to some extent because the income tax preference for agriculture is partly capitalized into the land price), small farmers are put at a disadvantage. Because their agricultural revenue will be insufficient to pay for the capitalized value of both the agricultural revenue and the tax benefit, they cannot benefit from the tax treatment and cannot buy land in areas with well-integrated land markets. Poor farmers are therefore forced to migrate to the frontier, where they clear new land from the forest.

Tax credits are provided (at a fiscal cost exceeding $1 billion between 1975 and 1986) to livestock ranches in the Amazon, thereby providing an incentive for clearing vast areas of forest and greatly reducing their biological diversity; some four million ha have been cleared to date using this incentive, even though most of the livestock ranches have a negative economic return. Similar credits are provided for reforestation schemes, but progress in reforestation has been very modest.

Regulations on the allocation of public land provide strong incentives for rapid deforestation to solidify claims on land and increase the size of final land allocation during the process of land adjudication. Ranchers with tenuous claims can be allocated two-to-three times the amount of land cleared of forest and put under pasture, up to a ceiling of 3000 ha per rancher. Land clearing also provides excellent protection against competing claims and against land invasions.

The World Bank report concluded that rural land ownership by non-farmers is much more common in Brazil than most other places in the world. However, Brazilian taxation, credit, and land policies provide additional strong incentives for investing in land and for acquiring it by deforestation. This subsidized deforestation is reducing biological resources at a rate which far exceeds the benefits being returned to the government or the people of Brazil; nor have the incentives have not been effective in creating viable livestock enterprises in the region.

(Source: Binswanger, 1987)

CASE STUDY 2: **ECONOMIC INCENTIVES RESULTING IN OVER-USE OF GRAZING LANDS IN BOTSWANA**

A combination of incentives has made the overstocking of grazing land in Botswana a response that is privately rational, and socially expensive. Livestock prices are most strongly influenced by the artificially elevated prices offered by the EEC, the major external market for beef. Increasing in real income terms over the past decade, they provide a strong incentive to expand livestock holdings (particularly as they rest on political agreements—the Lome Convention and the Common Agricultural Policy of the EEC—rather than international market conditions). When drought hit the country, the Botswana Meat Commission (BMC), which fixes prices for beef, paid high prices to provide short-term gains for livestock sellers, but instead of stimulating sales and reducing stocking rates this "bonus" perversely provided a direct incentive to increase stocking rates. BMC has also set the lowest prices at the onset of the dry season, thereby providing a disincentive to farmers to sell off excess stock during periods when the range is under hughest ecological stress.

In addition, deductability of capital expenditures stimulates investment in the livestock sector; livestock owners are provided with essential services are provided at low cost, including veterinary services, veterinary cordon fences, development of bore holes to provide water to cattle, and improvements to trek routes; and land rents are very low on tribal lands, making them attractive to cattle grazing.

These factors have stimulated the increase in the national cattle herd to levels that exceed the carrying capacity of the range. As a result:
- rangeland degradation is severe in a number of areas due to the combined effects of soil erosion, depletion of soil nutrients, and increasing soil aridity;
- the biomass and diversity of fauna and flora have been reduced in many parts of the country;
- in the wetter eastern areas useable rangeland is steadily declining, while the drier areas suffer from widespread devegetation, leading to reduction in organic and moisture content and to increased erosion, and ultimately to desertification;

- the availability and quality of water has been affected through increased run-off and sedimentation, leading to lower rates of recharge of groundwater, water losses in irrigation, reduction in surface water for wildlife, silting of dams, output losses in dam and river fisheries, and polluted drinking water.

The short-term gains to relatively few ranchers in Botswana has sent significant amounts of beef to Europe, at the cost of the long-term productivity of the biological resources in Botswana's arid lands.

(Source: Perrings, *et al.* 1988)

CASE STUDY 3: **INCENTIVES RESULTING IN OVER-EXPLOITATION OF TROPICAL FORESTS IN INDONESIA**

Governments with major timber resources often offer incentives which generate rapid logging by concession holders. These incentives include: limiting agreements to periods shorter than a single forest rotation (thereby providing no encouragement to replant); charging concession holders reforestation fees which are less than the cost of replanting; basing forest charges on the volume of timber removed rather than the volume of merchantable timber available (thereby encouraging only the most valuable trees to be taken, requiring a larger area to be logged to meet timber demand); and charging flat fees per cubic meter harvested, rather than adjusting the fee to the species taken (thereby providing a powerful incentive to take only the most valuable species).

Governments often further enhance the profits of the concession holders through additional indirect incentives, including support for international marketing, construction of roads and port facilities, and the costs of surveying, marking, and grading logs and timber for export. Further, governments also assume responsibility for the externalities, especially the loss of biological diversity to the nation. Other incentives are designed to enhance local wood-based industries, thereby increasing local employment and income. These include reduced or waived export taxes, disincentives against export of unprocessed logs, rebates on income tax liabilities, and long-term loans at favorable interest rates.

However, the economic costs in terms of lost revenues and faster deforestation can often be considerable. In Indonesia, for example, from 1979 to 1982, the total economic rents (generally speaking, profits) generated by logging for export approached $5 billion, but the official government revenue was just $1.6 billion. While some $500 million of potential government profits were lost because of inefficient domestic processing, some $700 million per year went to the private concession holders. It was hardly surprising that by 1983, the total area under concession agreements exceeded the total area of production forests in the country by some 1.4 million hectares. A series of such incentives increased the number of sawmills and plywood mills from 16 in 1977 to 182 in 1983, requiring an annual harvest from the forests some 50 percent greater than the maximum reached in the 1970s, when log exports were at their maximum. Worse, inefficient processing absorbed the rents available from the forests, with negative rents made feasible only by the government's financial incentives. It must be concluded that incentives to promote local processing do not necessarily contribute to conserving biological resources or diversity.

Repetto (1987a) concludes: "Overly generous logging agreements that leave most of the rents from logging virgin forests to concessionaires, and excessive incentives to forest-product industries that encourage inefficient investment in wood-processing capacity, combine to increase the log harvest much beyond what it would be without these policies. Poorly drafted and enforced forestry stipulations are inadequate to ensure sustainable forestry practices in the face of these powerful incentives."

(Source: Repetto, 1987a).

CASE STUDY 4: **USING ECONOMIC INCENTIVES TO INTEGRATE CONSERVATION AND RURAL DEVELOPMENT IN THAILAND'S KHAO YAI NATIONAL PARK**

Changing the behavior of local people toward biological resources of national concern usually requires a package of direct and indirect incentives, in cash and in kind. Nowhere is this better illustrated than at the village of Ban Sap Tai (population 500), adjacent to Thailand's oldest national park, Khao Yai. This village was typical of the remote region, with a low level of income, few community services, heavy debt burdens, a poor level of education, and many farmers without title to the land they farmed. As expected in such conditions, Ban Sap Tai was also notorious for its poaching of park wildlife, and village farmers constantly encroached on the park.

To solve the problem, an ongoing pilot project was initiated in 1985 to use creative rural development techniques in tandem with a conservation awareness program to encourage local cooperation in protection of park resources. The first element was a trekking program which was designed to link economic benefits for the village to the conservation of park resources. Villagers were used as guides and porters for groups of 10–12 tourists, spending several days hiking through park mountains above Ban Sap Tai; wages were US$5/day, three times the average rate for typical village labor. It was emphasized that preservation of wildlife and forests would encourage increased tourism, which in turn would bring outside money to the village and thus provide a direct incentive for villagers to conserve park resources.

However, it soon became apparent that the trekking program was not providing suffcient economic benefits to offer a viable alternative to illegal use of park resources, so additional incentives were added to the package. Expertise was sought from the Population and Community Development Association (PDA), an experienced development NGO. With funds from Agro Action, a German development NGO, PDA established an "Environmental Protection Society" (EPS), a unique, indigenous, community-based NGO—part credit cooperative, part non-formal education centre and part collective business enterprise. EPS membership is open to all villagers who possess proper land title and pledge to refrain from breaking park laws. An annual election is held from among the membership to elect a seven-person EPS committee which administers the EPS with assistance and training from PDA. The EPS also includes a Youth Group and a School Group.

The EPS's major goal is to act as a catalyst for income-generating projects, mainly agricultural, through serving as a credit cooperative. A revolving fund was established with $24,000 provided by Agro Action. EPS members borrow money at one percent monthy interest (compared to five percent interest offered by local middlemen) to purchase seeds, fertilizer and other essentials. Loans are recovered by the EPS after the produce grown with the loan money has been sold, with all repayments placed in the EPS revolving fund to be used in the next round of lending.

A cooperative store, operated by EPS members, was established to sell everyday goods at reasonable cost and generates additional revenue for the EPS. EPS members are offered shares in the store and receive regular dividends. Community woodlots were introduced to the village, satisfying both economic and conservation goals. The project also runs a "food for work" program whereby villagers perform community development work in exchange for rice; this is especially popular during times of the year when household rice stocks are low.

The project's major conservation component involves awareness and extension activities, both for adults and for children. Khao Yai Park staff offer periodic conservation awareness sessions to EPS members, emphasizing the many linkages of development with the proper care and maintenance of the natural environment. Students participate in twice-monthly lecture and demonstration programs on natural resources and conservation. The EPS and Khao Yai Park staff have cooperated in a tree-planting program to demarcate the park boundary, especially in those areas previously encroached by farmers. EPS members have also helped park staff to reforest park lands previously under illegal cultivation.

Training is an integral part of all project activities, so villagers become increasingly independent. EPS committee members are provided periodic training in management and administration of the revolving fund. In addition, members of the Youth Group are given basic training in management of cooperatives so they will be prepared to take over from their elders; such training also has practical application in management of various household businesses.

Strong emphasis is placed on technical training in improved cultivation techniques, handicraft production, conservation of forests, soil and wildife, and basic business skills, thereby optimizing productivity of the land, labor, and capital available in the village and reducing pressure to expand cultivation into the park. Several villagers have been trained as Village Health Volunteers, to provide basic health and family planning services.

The results of the project have been outstanding. EPS membership has grown from an initial 26 percent of all villagers in 1985 to over 70 percent in 1987. Moreover, 85 percent of all village households have participated in EPS functions. In 1986, the EPS revolving fund loaned a total of $23,000 to 73 members for crop production, to 35 members for cattle and chicken raising, to the Youth Group for soybean cultivation, to the school lunch program for maize production, and to the cooperative store for working capital. All loans were repaid in full. Profits from the 18 treks over the past two years have averaged approximately $200 or a total of US$ 3,600, and villagers are eager to serve as guides and porters in the program, viewing it as a useful income supplement.

The project has virtually halted encroachment on park lands; existing farms inside the park boundary were removed and no new plots were established. Further, creeping agricultural encroachment along the edge of the park has been virtually halted through cooperation between park officials and the EPS members. The location of the park boundary is now well understood and acknowledged by the villagers. Poaching has also been greatly reduced, and barking deer, elephant and other animals have returned to the edge of the village, a phenomenon not witnessed in over a decade according to long-time residents.

Agro Action, the original funding agency, has been so impressed with project results that they have provided full financial support for expansion to two more villages adjacent to Ban Sap Tai, and other agencies have established similar projects in eight additional villages along the Khao Yai boundary.

(Source: Praween, Tavatchai, and Dobias, 1988)

CASE STUDY 5: **ACCESS TO THATCHING GRASS AS AN INCENTIVE FOR PROMOTING LOCAL SUPPORT FOR PROTECTED AREA VALUES IN ZIMBABWE**

Zimbabwe's Matobo National Park is threatened along most of its boundaries by dense settlement of pastoralists in degraded habitats. Because their own lands are seriously over-grazed, the villagers see the lush grazing in the park as an important resource which should be available to their cattle; further, some of the villagers resided within the present park boundaries until the mid-1950s and still consider the park to be "theirs."

Thatch is the main roofing material in this part of Zimbabwe but is now in extremely short supply due to overgrazing. But within the park, thatching grass occurs in such quantities that park managers burned it periodically to enhance grazing for large wild mammals (the primary management objective for the park) and to prevent a build-up of dead grass which could feed a catastrophic conflagration.

In 1962, the park authorities met with the local communities and agreed that instead of burning the thatch grass, they would permit the villagers to harvest it under a strictly-controlled regime. In order to use this privilege as an incentive to control poaching of wildlife, trespassing with cattle (which might carry disease to the wildlife), and setting of fires, park authorities traded thatch collection for an understanding that the local people would strictly abide by the protective legislation. Village elders nominate villagers, mostly women, who are licensed to cut a given number of bundles of thatch according to a pre-determined annual quota and "pay" the park authorities one bundle for every ten cut (on the principle that "free" goods are not valued by the recipient). The park's share of the grass is used to roof visitor facilities and service buildings in the park, thereby also bringing a tangible benefit to the protected area.

Annual quotas have ranged from around 40,000 to 115,000 bundles, bringing an income of $20,000 to $60,000 to the community. Poaching and wild fires have been minimized and cattle trespassing is much less serious than might otherwise be the case in a politically sensitive region. Other benefits include reduced pasture management costs, the protection and rehabilitation of important water-generating catchments, the availability of a steady supply of attractive cheap roofing material to the park, and improved relations with neighboring villages.

(Source: MacKinnon *et al.*, 1986)

CASE STUDY 6: ECONOMIC INCENTIVES FOR CONSERVING BRAZIL'S IGUAPE-CANANEIA-PARANGUA ESTUARY

In 1985, the state governments of Sao Paulo and Parana states identified this estuarine area as an important ecosystem to be protected for the benefit of the local communities. The effort involves: reinforcement of protection measures for the existing parks; sustainable use of biological resources; land use regulations; improved management of marine resources; improvement of health and sanitation services; integrated ecosystem research; and environmental education.

The core of the project is a coastal zone management plan that will indicate what economic incentives are to be used to achieve the objectives established for the project. In preparing the plan, conflicting uses will be resolved by a Coastal Committee formed by representatives of government institutions, fishermen and peasant associations, entrepreneurs, and environmental groups.

A management plan was completed in 1986 for Ilha Comprida, a sand barrier island under threat by intensive tourism. Priority was given to land use control, conservation of mangroves, and mariculture. Along with this management plan and reinforcement measures for existing state parks, the program has initiated a set of economic projects aiming at improving the living conditions of the local population and encouraging their cooperation with park objectives. These projects are designed to use a variety of renewable resources of the estuarine area within the framework provided by the traditional economic system based on a mixture of agriculture, fishing, and other activities. Some examples of these projects are:

Oyster culture, combined with small-scale agriculture and fishing. As the mangrove oysters from Cananeia are being depleted, the Secretary for Environment of Sao Paulo and the Fisheries Institute are starting a community-based project of oyster culture, on a community basis. The scientific know-how is already available through years of biological research undertaken by the Institute. Economic incentives will be used to maintain the complementarity of traditional economic activities. Transportation in support of agriculture, fishing and oyster production will be subsidized.

Processing of local products. Some projects are being undertaken in order to increase the local profit from local products such as fish and agricultural products. Fish smoking is a technique which is being introduced in the area in order to increase the value of fish resources. The idea is to raise the income of fishermen without a higher fishing effort, thus reducing pressure upon the resource base.

Palm-tree plantation. There is strong pressure on palms *(Euterpe edulis)* for collection of palm hearts. Although the cutting of the trees is forbidden, some local communities depend on this activity for their livelihood. Recent research has shown that cultivation of *Euterpe* is feasible in the forest. Incentives are being considered in order to plant these trees in their natural environment.

(Source: Diegues, 1987)

CASE STUDY 7: **INCENTIVES FOR CONTROLLING HUMAN IMPACT ON THE FORESTS OF SAGARMATHA NATIONAL PARK, NEPAL**

Some 2500 of Nepal's estimated 20,000 Sherpa people live in the 124,000 ha of Sagarmatha National Park (which also contains Mt. Everest). The Sherpas had traditionally managed their mountain habitat through trading and limited agriculture, and had a relatively benign impact on the environment. With the nationalization of the forests and the coming of mountaineering and tourism in the 1950s, all this changed. The forests were over-exploited, leading to erosion and shortages of domestic firewood; perhaps less tangibly, the Sherpa culture was affected by the influx of tourists.

When Sagarmatha National Park was established in 1976, the initial reaction of many Sherpas was one of hostility. But they have been won over by a series of incentives aimed at bringing the benefits of the protected area to the Sherpas. These incentives include:

- major employment opportunities in tourism, as porters, trek leaders, and hotel owner/managers;
- preferential employment as National Park staff (nine of whom are Sherpas);
- registration of land to establish tenure rights;
- returning responsibility for forest protection to the community, including providing financial incentives to local elected forest guards *(Shingo nawa);*
- restricting the use of firewood to residents, and requiring hotels to use kerosene;
- overseas training opportunities for local park staff (several have been provided scholarships to New Zealand, the UK, and USA);
- restoration and protection of religious structures within the park, and prohibition of trekking and mountaineering in sacred areas, and on sacred mountains (including the 6856 m Ama Dablam);
- community development, including mini-hydropower and solar power systems and improved insulation of Sherpa dwellings.

The objective of these incentives is to stimulate the recovery of the forests of the mountains surrounding the villages and to revive important elements of Sherpa culture.

(Source: Jefferies, 1985; Norbu, 1987)

CASE STUDY 8: **ACCESS TO GRAZING AND WATER AS AN INCENTIVE TO CONSERVE THE AMBOSELI ECOSYSTEM, KENYA**

Kenya's Amboseli ecosystem typifies the problems of conserving large mammal communities in Africa. Amboseli's wildlife migrates seasonally beyond the confines of the park boundaries—in this case onto land owned by Masai pastoralists. Traditionally the Masai were subsistence herders, but as their lifestyle changed to a more settled existence they became increasingly unwilling to accept wildlife on their lands since the animals contributed nothing to the local human economy, even though the value of wildlife nationally through tourism was considerable.

The Amboseli basin, fed by permanent springs from nearby Mt. Kilimanjaro, is the only source of permanent water in the region, a source of conflict between Masai cattle and wildlife. The Masai sought land tenure to the entire region including the Amboseli basin. They argued that revenues from tourism went only to the Kajiado County Council 150 km away and contributed nothing to the local economy, and that wildlife had traditionally served as the Masai's "second cattle" during droughts but now hunting was banned. Why then should the Masai lose their traditional dry season grazing grounds to benefit the Government, the Council and the tourist?

The dispersal of wildlife over some 5000 sq km during the rainy season created insurmountable obstacles to conserving the whole ecosystem; some 6000 Masai, 48,000 cattle and 18,000 sheep and goats depended on the same area and could not be relocated elsewhere. Over 80 per cent of the wildlife migrants concentrated each dry season around the 600 sq km of the basin but this area was inadequate as a self-sustaining national park, as the large herbivore population would decline by 40-50 percent if confined permanently to the basin. Similarly, if the Masai were deprived of the basin's water and swamps their livestock would decline by half.

A package of incentives was used to promote a compromise between the Amboseli National Park and the local Masai. A water diversion scheme was built to pipe water from the springs to artificial swamps created outside the park for the Masai herds, thereby removing the domestic cattle from the park. To provide the wildlife migrants with needed access to Masai lands, a plan was agreed whereby in return for continued access to the entire ecosystem Amboseli's wildlife, the park would pay a grazing compensation fee (to cover their livestock losses to wildlife migrants), Masai would control hunting and cropping on their land, and would be provided subsidies to enable them to accommodate tourist campsites and lodges. The net monetary gain of the park per year from continued use of the Masai lands would be approximately $500,000 and the benefits from the park to the Masai would ensure them an income 85 per cent greater than they could obtain from livestock alone after full commercial development. The new park headquarters is located in the south-eastern corner of the park and includes a local community centre with a school and medical facilities.

The park has become a source of employment, revenue and social services.

(Source: Western, 1984)

CASE STUDY 9: "OWNERSHIP" OF MARINE RESOURCES IN QUINTANA ROO, MEXICO

Fisheries have long provided important biological resources for supporting human communities, but the Yucatan Peninsula of Mexico is unique in having a period of 400 years (1500–1900 A.D.) when the coastal zone was depopulated; when villages again became established early this century, new means of managing marine resources needed to be developed. Clear "ownership" has proven to be an important incentive for effective management of an economically important resource, the spiny lobster (Panulirus argus). International support in the form of recognition of the area as part of a Biosphere Reserve has helped ensure that the new management system is sustained.

Large-scale commercial fishing did not begin in this part of Mexico until the mid-1950s, focussing on the production of high-cost export species such as lobster, conch, and shrimp. Numerous cooperatives were formed in order to comply with Federal law which reserved these species to cooperatives, and each was provided a site where they could fish. However, only two co-ops have proven successful, the others having failed largely because they did not live up to their objectives; anyone could harvest anywhere within the co-op territory, and exclusive rights were not given to the individuals who established habitat improvement measures.

The two successful cooperatives occupy the Ascencion and Espiritu Santo bays, and have improved the habitat for lobsters by providing shelters that simulate attractive natural features. Each fisherman within the cooperative harvests only from his own territory, where he establishes and maintains his own lobster shelters. By common accord, the fishermen do not place the shelters within 25 meters of their territorial boundary. Territories can be bought, sold, and traded among cooperative members, with prices determined by potential for profit; while formal titles do not exist, the territories are sufficiently well recognized that they can be inherited by a spouse or divided among children.

"Ownership" of a territory transfers considerable control over access to most other marine creatures within it. The cooperatives have agreed certain limitations, such as closed seasons and permissible equipment. Members of the cooperative are careful to avoid even the suggestion of improper use of a territory belonging to another member. Fishermen are very conscious of their responsibility for protecting their territory, and do not hesitate to inspect the boat of any fisherman found in their territory. Penalties for poaching are severe, including banning from the cooperative and confiscation of equipment by the cooperative (note that government Fisheries Department officials lack adequate resources to effectively police the area).

Ascencion and Espiritu Santo bays are located in the new Sian Kaan Biosphere Reserve, which is dedicated to the sustainable use of biological resources. The combination of incentives such as use rights and peer support and disincentives such as confiscation and peer pressure have led to a management system that works.

(Source: Miller, 1986)

CASE STUDY 10: **FOOD FOR WORK AS AN INCENTIVE FOR COMMUNITY ACTION IN WOLONG NATURE RESERVE, CHINA**

An excellent example of food-for-work was a World Food Programme project on "Development and Protection of Wolong Nature Reserve" in Sichuan Province, China (a biosphere reserve and important habitat of the giant panda). In responding to an emergency situation where pandas were starving because their bamboo dietary staple had died, the project was designed to remove human pressure in an important natural area, and to encourage the pandas to move down to lower ranges of the reserve. Some $770,000 worth of food was provided to some 3400 local people in the form of 7 million per capita rations over the period of one year, to carry out the following activities:

- building new houses to resettle 100 households from the fragile uplands of the reserve (close to the habitat of the endangered giant panda) to the more productive lowlands; no rent was charged for the new houses;
- Building a school for 400 pupils;
- Contributing to the construction of a 600 kilowatt hydroelectric power station which was designed to provide an alternative source of energy to the use of firewood harvested from the Wolong Reserve;
- Constructing some 100 km of footpaths, in order to facilitate patrolling of the reserve;
- Planting some 1000 ha of previously cultivated and abandoned land with bamboo varieties known to be favored by the panda; and
- Patrolling the reserve, using teams of 10 workers selected on the basis of their experience and knowledge of the reserve, in order to locate starving pandas and to provide emergency panda food in strategic areas.

A number of elements combined to make this a reasonably successful project. The biological resource under threat—the giant panda and its ecosystem—was of great concern to the government; the solution to the problem of conserving pandas and panda habitats could be addressed at least partly through labor; the area was relatively poor, and suffered from a food deficit so that food aid was an effective incentive; an effective government infrastructure existed; training programs were provided for panda protection staff; WWF provided additional assistance for protection of the panda, in the form of equipment and expertise; sufficient labor was available; and government support was available to cover the costs of materials, supplies, distribution of food, and wages of government workers.

As of April 1984, World Food Programme commitments to China had amounted to $194 million, indicating the potential influence of food-for-work projects. Similar efforts could be undertaken in other food-deficit areas, especially in Africa, where food can be exchanged for work in support of conserving biological resources.

CASE STUDY 11: **PROVIDING PROFITS FROM HUNTING TO LOCAL COMMUNITIES: A MAJOR INCENTIVE FOR CONSERVING LARGE GAME IN ZIMBABWE**

The proper use of wildlife offers one of the best opportunities for redressing the socio-economic and environmental plight of much of the drier parts of Africa. Since local people are likely to be the best managers of wildlife on their lands, the Government of Zimbabwe has enacted legislation to give landholders the rights to use wildlife, other than a limited number of Specially Protected Species, while it was on their land. State hunting licenses were abolished in favor of those issued by the landholders who could charge for them as they pleased, thereby gaining a significant economic incentive for conserving their wildlife. Mechanisms were, however, provided whereby abuses of these rights could be controlled by the local landholder community, or by the State if necessary.

Such management made sound economic sense. In the more productive ranching areas, receiving about 600 to 800 mm per year of rainfall, the profits from cattle ranching can be raised 50 percent or more by introducing a complementary wildlife enterprise on the same land. In areas with rainfall of 450 mm or less, the gross return per unit area on a sample of ranches where wildlife was well managed was some four times greater than that from well-conducted cattle ranching.

To promote the conservation of the wildlife resources found on communal lands, "private game reserves" have been established where revenues from hunting would be paid to local communities instead of into the consolidated revenue fund. Through this arrangement, some $4.5 million has been paid out for development in remote parts of the communal lands over the past seven years.

To promote the conservation of communally-owned biological resources, a series of agreements are being concluded with small social units, formed into public companies, in which every adult has an equal share, assuring each member an equitable return from the resources. Entry into the program by a community is voluntary, with the intention that the community would eventually assume control of its own affairs. This would meet four essential objectives:

- to allocate rights to use and benefit from resources to an identifiable group of people who could be held accountable for proper management of the resources;

- to give individuals a democratic voice in the corporate management of their resources and a personal choice in the use of the benefits from them;
- to provide an individual and corporate incentive to invest in the protection of their life support systems;
- to provide resources of growing scarcity with a tangible monetary value.

Recreational hunting is now the most positive and widespread economic incentive for the conservation of large mammals in Zimbabwe.

(Source: Child, 1988a)

CASE STUDY 12: **COMMUNITY INVOLVEMENT AS AN INCENTIVE FOR CONSERVATION MANAGEMENT OF A WOODLAND IN THE INNER DELTA OF THE NIGER, MALI**

In the Inner Delta of the Niger River in Mali, woodlands of Acacia kirkii require seasonal inundation to ensure their growth cycle, and the flooded thorny woodlands which result are essential to the breeding success of colonial waterbirds such as herons and cormorants. In turn, feces and regurgitated food from the breeding colonies fertilize the waters and support an economically important fishery. Local farmers also benefit from the flocks of cattle egrets which eat millions of grasshoppers daily, making a significant contribution to crop protection in nearby millet fields.

The traditional land use system within the delta divided the area into discrete fishing grounds managed by villages during the flood season. When the same areas dried out, the herding communities, represented by a powerful individual called the Dioro, administered the pastures, controlling access to herders.

With independence in 1960, the Government of Mali nationalized all land and the management of fishing, grazing and woodland exploitation was put in the hands of the Ministry of Natural Resources, thereby greatly weakening traditional systems of resource management which had functioned for hundreds of years.

Today a major problem concerns the goat herders who migrate into the delta area during the dry season. They buy a Cutting Permit at the Forestry Department which gives them permission to construct a thorn enclosure where the goats spend the night. Fines are levied on those who cut live trees to feed their goats, but collective fines are imposed as the culprit is rarely caught red-handed. All herders in the area contribute to the fine; so the rational herder decides to cut trees since he will have to pay anyway. The disincentive of fining appears to the herders to be entirely independent of their actions, thereby invalidating the intent of the disincentive.

The result is predictable: Tree-cutting to give goats access to foliage has reached levels likely to damage the future of the woodland both for goat grazing and for the waterfowl colonies (and hence for part of the fishery). An IUCN project, with funding from a number of development assistance agencies, is seeking a solution through supporting Forestry Department efforts to create "Forests

Villageoises," which are managed by local committees consisting of goat herders, fishermen, and Dioros.

An essential element has been the retention of the traditional control by Dioros over grazing areas. The ownership of the forest is vested in the local village, which has interests primarily in fishing and is willing to recognize the traditional authority of the Dioros. The number of goat herds is reduced in one case to 20, a move which is welcomed by the goat herders. By re-creating traditional control structures, woodlands in the Inner Delta are now being conserved for the benefit of fisheries, waterfowl, and grazing.

(Source: Skinner, 1987)

CASE STUDY 13: **PROVIDING TRAINING TO FARMERS AS AN INCENTIVE FOR CONSERVING THE TEGUCIGALPA WATERSHED, HONDURAS**

The 7500 ha La Tigra National Park provides over 40 percent of the water supply for Tegucigalpa, the capital of Honduras. The park is surrounded by an officially-declared buffer zone of 14,500 ha, most of which is privately owned and is being exploited in ways that detract from the watershed protection function of La Tigra. In order to develop more effective ways of protecting the watershed and its biological resources while bringing enhanced benefits to local people, a multidisciplinary team from 10 government institutions and from several of the communities and agricultural cooperatives in the buffer zone designed a new approach to protection.

The Operational Plan prepared for the area called for strengthening the capacity of park staff to enforce regulations. Providing infrastructure such as guard posts, patrolling trails, uniforms, and equipment enabled the existing legal disincentives to be implemented.

More important were the incentives provided to the farmers in the buffer zone. Based on a process of socio-economic studies and consultation with local farmers, several pilot rural development projects were designed to maximize self-help, confidence building and control over resources by the villagers themselves, with minimum investment costs and heavy emphasis on labor-intensive, integrated management technologies.

A critical element was the establishment of a few training farms within the buffer zone. Training covers techniques of all key types: agroforestry practices; multiple cropping; crop rotation; biological control methods for pest management and integrated pest management; firewood production; home-made tool making; use of farm wastes as animal food; management of cooperatives; marketing skills; etc. The training is conducted almost entirely as practical hands-on exercises, with the participants directly working on the demonstration farms. Following the week-long basic course in integrated farm management, the participants receive regular visits for several years to their home farms, helping them to put into practice their new knowledge and techniques.

Managed by IUCN, the program is being funded by governmental and international agencies (including NORAD, CIDA, and

WWF). It is expected that the demonstration farms will soon be-come self-sufficient, through sales of farm products; one of the farms is already earning $500 net profit per week, after operational ex-penses. The objective of this effort is to encourage the Honduran Government to institute mechanisms that will eventually internal-ize costs for the effective management of La Tigra National Park and its buffer zone, through such means as a tariff system for water services (with part of the earnings being devoted to La Tigra) and creation of a trust fund as part of each major water development project based on water from La Tigra.

CASE STUDY 14: **TRADITONAL OWNERSHIP FOR MODERN CONDITIONS IN THE COASTAL ZONE OF JAPAN**

In contrast with the maritime tradition prevalent in the Western world, Japan has never adopted the idea that the sea is a common property resource, owned at once by everybody and nobody. On the contrary, over many centuries a complex system developed which provided various forms of customary village tenure and rights to fisheries in coastal marine waters. These traditions were so effective in preventing abuse of the resource, that they have been incorporated into national legislation.

The local Fisheries Cooperative Association (FCA) is the main corporate fisheries rights holding group. Each FCA belongs entirely to a local community of fishermen, and the FCAs hold fishing rights to virtually all coastal waters. These rights continue historical practices and protect coastal fishermen against other fisheries and economic sectors by granting them fully protected property rights. However, while these rights are regarded as the exclusive property of the fisherman to whom they are granted, they cannot be loaned, rented, or transferred to others.

FCAs determine the division of access rights among individual members of the cooperative and ensure that all interests are accounted for. They also permit fishery regulations instituted by government fisheries agencies to be adapted to regional differences in ecology, target species, fishing effort and level of industrialization, and ensure that management strategies, processes of conflict resolution, and inter-personal and inter-group relationships will be based on local customary law and codes of conduct.

FCAs adapt to local conditions. In Hokkaido, gill-netters may fish anywhere in the FCA's territory but fishermen operating small-scale fixed nets are regularly assigned the same fishing spots because the nets must be individually tailored to the bottom topography in each spot fished. In other regions, octopus holes within a joint rights area are owned and inherited as personal property; lotteries are used to allocate valuable fishing spots among FCA members; or free competition and first-comer's rights may prevail (especially in less productive locations). Although legally all fisheries rights waters belong to all members of an FCA, in practice small spots within such a sea area are conceived of as temporarily belonging to an individual fishing unit. This private "ownership" within the common domain promotes equitable access to resources, minimizes interpersonal conflict among fishing units, and avoids over-fishing of any one area.

The drawback to this complex combination of the traditional and the modern is that it makes comprehensive coastal zone planning extremely complicated, and indeed almost impossible. Nevertheless, Japan provides a well-functioning example of how common property resources can be managed effectively.

(Source: Ruddle, 1986)

CASE STUDY 15: ECONOMIC INCENTIVES AMONG PASTORALISTS IN NORTHERN KENYA

The Mt. Kulal Biosphere Reserve covers over 7,000 sq km of the arid and semi-arid zone of northern Kenya. It is the home of four tribes of nomadic pastoralists—Rendille, Gabbra, Samburu, and Turkana—who keep camels, cattle, sheep and goats. Although they previously lived in reasonable balance with their environment, they are now threatened by frequent droughts and the associated loss of vegetation cover and soil cover resulting from high human and livestock population pressure.

To promote a new balance between people and resources, the Integrated Project on Arid Lands (IPAL) devised a series of economic incentives and disincentives which were aimed at conserving land, wildlife, and local cultures. These incentives were incorporated in a series of resource management guidelines on the use of water, wildlife, grazing resources, woodlands, water catchments, soils, fisheries resources, livestock, and human resources. The guidelines were prepared in full consultation with the local people, and operate within the constraints imposed by the traditional pastoral economy of the tribal people. Incentives in the project included:

- registering tribal rangelands in order to put them on a firm legal basis;
- providing subsidies for the development of water resources, marketing facilities for livestock, and banking facilities to store wealth other than "on the hoof";
- providing security against raids from other tribes (such raids prevent about 40 percent of the area from being used);
- providing conservation education in schools, wildlife extension in adult literacy classes, and information for government officials about the value of conservation;
- providing employment for local people in the system of protected areas; and
- providing income from tourism to the protected areas for development activities such as health and water development.

Disincentives were designed to prevent grazing on steep slopes, to control the stocking rates of livestock, and to enforce a moratorium on grazing certain pastures in poor condition.

By improving the productivity of the best grazing areas, the marginal lands were given improved protection and representative examples of the original flora and fauna were conserved. Protected status was given to the forests on Kulal, Olsonyo, Mara, and Marsabit mountains in several management categories (National Reserve, National Park, and Biosphere Reserve) which allowed various degrees of human use. Traditional "drought reserve" rangelands were included in the protected area system, to be used for grazing only in drought emergencies.

(Source: Lusigi, 1984)

CASE STUDY 16: **INCENTIVES AND DISINCENTIVES IN COMMUNITY-LEVEL MARINE RESOURCE MANAGEMENT IN THE PHILIPPINES**

Villages in the coastal zone of the Philippines have long been dependent on the productivity of coral reefs, but as traditional management systems break down, overexploitation has increased. A project carried out by Silliman University, USAID and the Asia Foundation aimed at enhancing fisheries resources through building new systems of responsibility for resource management in three island villages, each having a coral reef small enough to patrol.

Accepting that the only effective resource management would come at the community level, project staff encouraged individuals interested in the problem of marine conservation to form Marine Management Committees (MMC). The MMCs matured into working groups which received community respect, once the entire community decided to implement a marine reserve management scheme, a process involving give-and-take among project staff, local officials, and residents.

As the marine reserve began to function and illegal fishermen were repelled, the community gave more support. Building an education center with local participation and supervision provided a source of community pride. MMCs became more effective as they were given new responsibilities for projects such as placing Tridacna clams in the fish sanctuary areas for the community to manage and harvest; refining the marine reserve guidelines into a legal document adopted by the municipal town councils; training MMC members in the management of tourists to the coral reefs; developing education programs for all parts of the community; and initiating alternative income schemes such as mat weaving and sea cucumber mariculture.

Three marine reserves with municipal legal support are now demarcated by buoys and signs, and managed by MMCs which actively patrol for rule infractions by local residents or outsiders. Copies of the municipal ordinances are posted on the islands in the local language and published in a brochure. Each site now has a fishery breeding sanctuary and a surrounding buffer area for ecologically-sound fishing. Destructive fishing methods—such as using dynamite, cyanide or other strong poisons, and very small mesh gill nets—which were formerly widespread have now been effectively banned. Increasing numbers of tourists are visiting the sanctuaries, and bringing economic benefits to the villages.

Species diversity and abundance have significantly increased for certain families of fish, especially the favorite targets of fishermen; mean percentage increases in species diversity ranged from 25 to 40 percent, while increases in the numbers of all food fishes ranged from 42 percent to 293 percent over the three sites. In addition, and of crucial importance for the sustainability of the new reserves, the total fish yield for the fishermen has also increased; protection of part of the sea from fishing pressure has thus led to a total net increase in productivity and economic benefits for fishermen.

(Sources: Savina and White, 1986; White and Law, 1986)

CASE STUDY 17: **INNOVATIVE FUNDING FOR CONSERVING BIOLOGICAL RESOURCES IN COSTA RICA**

Costa Rica has Central America's most outstanding system of protected areas, covering over eight percent of the nation's territory. But current economic conditions have forced Costa Rica to seek extra-budgetary mechanisms for establishing and managing the system, involving a wide range of incentives and disincentives and the collaboration of numerous government agencies.

A large percentage of land acquisition costs for Costa Rican parks has come from the Agrarian Development Institute, which has issued special national parks bonds to expropriate many land holdings. Other areas have been purchased by the Agrarian Reform Institute as part of colonization projects but later deemed to have greater value for conservation.

Many other government agencies have provided manpower, equipment, and occasionally monetary support to park management. Examples include Public Works Ministry support in building and maintaining access roads; Public Security Ministry manpower support for dealing with protection programs; Tourism Institute financing of park infrastructure; Planning Ministry support for resource inventories and management plans; National Museum and university support for research programs; and National Youth Movement provision of volunteers.

The system is now well established, and the challenge has shifted to providing adequate protection to the parks and reserves and maximizing the long-term provision of goods and services from them. Special proprietary funds have proven to be a reliable source of revenue for operating expenditures of the conservation authorities, drawing on donations, transfers from other agencies, fees and charges for visitor services and concessions, and a series of fiscal stamps. Major sources of revenue, which totalled over $400,000 in 1987, include:

- *Fees charged park visitors and concessionaires,* which are expected to generate $168,000 in 1988. Concession fees for operation of a series of radio and television towers and a concession for a refreshment stand at the zoo are expected to generate another $35,000.
- *Fiscal stamps* created through legislation stipulating that all legal documents at the municipal level, newly issued passports, exit visas, first-time auto registrations, authenticated signatures registered at the Foreign Ministry, as well as all bars, nightclubs,

dance halls, and any other place that sells liquor, and all places of entertainment such as pool halls, cinemas, casinos, and public pools, must purchase fiscal stamps with at least part of the revenue being returned to the conservation fund. Additional fiscal stamps which contribute to conservation are required from new motor vehicle registrations, annual vehicle registrations, and wildlife import and export permits.

- *Hunting licenses* for small game and large game (with foreigners being charged more than locals), and fresh water fishing licences.
- *Excise taxes* on arms and ammunition and income from fiscal stamps. These are potentially important, but have declined drastically in recent years. The stamp prices were set by law in 1977 and have not been increased since then, because a new law would have to be passed by the legislature to vary the amounts. The Costa Rican colon is now worth only 11.4 percent of its dollar value in 1981, and this devaluation has been accompanied by rampant inflation. The dollar value of fiscal stamp receipts in 1982 was over $86,000, nearly three times that expected for 1988. Much of the 1987 revenue had to be used to pay for a new issue of the stamps.
- *Transfers from other government agencies.* Through a series of decrees establishing parks passed in the 1970's, it is mandated that the Tourism Institute must provide financial support to protected areas important for tourism.

These various proprietary funds have enabled systems of disincentives to be implemented, which in turn generate additional funds. The ready availability of funds for fuel, a major limitation in the past for game wardens, has made it easier to patrol large areas and has helped generate an increase in the number of individuals buying fishing and hunting licenses.

Establishing and operating the proprietary funds for the National Park Service, whose lands cannot be opened to extractive use, and the Wildlife Service, whose lands are legally designated for multiple use and therefore have more options for producing income, have required strong policy support from the central government. In addition to the enacting legislation, government has had to resist very strong pressure from the International Monetary Fund and other holders of Costa Rica's external debt who would like to see all proprietary funds eliminated. While most such funds have been eliminated by the Costa Rican legislature in recent years, strong lobbying by the local conservation community has to date staved off efforts to eliminate the Parks and Wildlife Funds.

(Source: Barborak, 1988a)

CASE STUDY 18: **THE RUBBER TAPPERS MOVEMENT IN BRAZIL: HARVEST RIGHTS AS AN INCENTIVE TO CONSERVE BIOLOGICAL RESOURCES**

Providing local resource users with the responsibility for using their resources sustainably can often be used as an incentive for conserving the larger ecosystem. An outstanding example is the emerging "rubber tappers movement" in the Amazonian region of Brazil, where some 500,000 people earn a living from collecting latex from wild rubber trees; the value of the forest products collected in the province of Acre in 1980 totalled some $26 million. In Acre, the per hectare value of extraction is more than twice that of cattle ranching, even without taking sustainability into consideration, and since 1970 the per hectare value of extraction has increased more than that of either agriculture or ranching.

However, the tappers do not have title to the forests they harvest, so they have organized a series of cooperatives aimed at gaining legal guarantees for maintaining their non-timber extractive uses of forest lands. The National Council of Rubber Tappers, founded in 1985, is therefore creating "extractive reserves"— protected areas to be sustainably managed by the communities that live in and know the forest. By granting use rights to the tappers, policy makers—with support from the World Bank and the Interamerican Development Bank—are protecting the forests against other uses and are thereby contributing to the conservation of biological resources; the sustainability of extraction, and the fact that it does not destroy forest, makes it a particularly attractive alternative to agriculture and cattle ranching.

Legally protected areas under local control are appealing to communities facing expulsion by cattle ranchers, large landowners, or colonization projects. Further, the security provided by legal protection of forest lands could be an incentive to increase production, making the proposal even more attractive economically.

The Acre Pro-Indian Commission (CPI) has helped extractive communities establish cooperatives and earn recognition for Indian land rights covering nearly 15,000 sq km, almost 10 percent of the land area. Given the legal guarantee to land rights that Indians hold under the Brazilian constitution, and the explicit will of the groups to preserve the mixed economy that they practice (small-scale agriculture, hunting, fishing, and rubber and Brazil nut gathering for cash income), the indigenous communities of Acre are an important constituency in support of forest protection.

The most important guarantee that reserves are defended will be creation of conditions for extractive production to compete effectively on the market, through increased productivity, improved marketing of extractive products, removal of subsidies for unsustainable land uses, access to credit for extractive producers, and improved health services and education. Policy support required from the national government includes appropriate pricing policies for rubber and legal mechanisms for the establishment of extractive reserves.

(Sources: Schwartzman, 1987; Allegretti and Schwartzman, 1986)

CASE STUDY 19: ECONOMIC INCENTIVES PROVIDED TO RURAL COMMUNITIES ADJACENT TO INDIAN WILDLIFE RESERVES

In 1983, the Task Force of the Indian Board for Wildlife recommended that government agencies should recognize the rural areas surrounding wildlife reserves as Special Areas for Eco-Development (SAED). This status was provided to these areas in recognition of their expected contribution to national objectives for conserving biological resources, and to compensate these villages for any sacrifices they might need to make. The following activities were suggested as appropriate incentives and disincentives to be applied in these Special Areas:

Forestry

- All forestry operations in buffer zones will include wildlife conservation as a major objective.
- The use of forests in buffer zones will be restricted to the local communities, and all such use should be sustainable.
- Soil conservation projects will be implemented in eroded areas.
- Pasture development and afforestation of denuded areas and forest management will be planned primarily to meet the pasture and firewood needs of local communities.
- Monocultures will be discouraged and efforts would be made to preserve and regenerate natural diversity in forests.

Agriculture

- Projects will be supported to develop and apply improved dry farming techniques for marginal lands, including improved seeds and fertiliser regimes.
- Cash crops will be permitted, even promoted, where they are likely to be more profitable than cereals, provided their cultivation is sustainable.
- Diversions, storage dams, and minor irrigation schemes will be supported.
- Soil conservation on forest and agricultural lands will be supported as a part of catchment treatment for major irrigation projects in lower reaches of major rivers.

Animal Husbandry

- The cattle population will be gradually reduced, and breeds will be improved through castration of scrub bulls and controlled breeding of healthy cows and buffaloes with bulls of good stock.
- Fodder farms will be established where feasible.
- Goat-keeping will be discouraged, and the government would not sponsor any goat-keeping program.

Tribal Welfare and Rural Development

- Local art and handicrafts will be promoted through sale outlets in tourist complexes of well-visited wildlife reserves.
- Local people will be given preference in employment.

(Source: Government of India, 1983)

CASE STUDY 20: **CREATING A REVOLVING FUND FOR SUPPORTING WILDLIFE MANAGEMENT IN ZAMBIA**

The Luangwa Valley of Zambia is one of the richest wildlife areas in the country, containing four national parks which cover about 20 percent and game management areas which cover about 60 percent of the valley . The game management areas differ from parks in that they are zoned for wildlife utilization and allow human occupation. However, legally all wildlife is the property of the State, and to harvest the animals in game management areas requires licenses that are often prohibitively expensive to residents.

The wildlife resources of the Lupanda region of the Valley has supported safari hunting which yields about $350,000 per year; but less than one percent of the safari hunting revenue was returned to support local village economies and a negligible amount went toward wildlife management costs. As a result, local support for conservation has been very low, and illegal hunting of wildlife, especially elephants and rhinos, reached such high levels that extirpation of some species was a real possibility. Despite strengthened law enforcement, villagers welcomed poachers from distant parts of the country, as long as they shared some of the harvested meat with the community.

Based on the premise that at least part of the revenues earned from wildlife should be returned to the National Parks and Wildlife Service (NPWS) to management the wildlife resource, a Wildlife Conservation Revolving Fund was established in 1983. This Fund also enabled the NPWS to employ additional staff beyond the Government approved civil servants.

Wildlife Sub-Committees were established in each Chiefdom, administered by a Unit Leader from the NPWS, and a village scout program was initiated. The scouts were given a training course of six months, officially designated as wildlife officers, and employed throughout the year in their respective chiefdoms as the local "custodians" of their village's wildlife resources. Additional manpower was also recruited from the local community on a seasonal basis to assist with other management needs, including building maintenance and construction.

Income to the Wildlife Revolving Fund came from the harvest of hippos, and from auctions among safari hunting companies for the rights to hunt in the Lower Lupande Game Management Area, with terms of the auction included quotas on animals which could be taken and minimum levels of employment from the local communities. Forty percent of the proceeds from the auction was handed over to the local Chiefs for community projects of their choosing and 60 percent was devoted to wildlife management costs.

Results have been remarkable. Manpower increased from 11 to 26 from 1985 to 1987, and the number of field-days by staff increased from 176 to 717 man-days. Annual mortality of elephant and black rhino, expressed as the number of poached carcasses found per year per square kilometer, decreased by 90 percent in the same period. In 1987, the total earnings for the Revolving Fund were $48,620, of which $14,840 was devoted to wildlife management, including $4,410 for the village scout program. Overall recurrent costs of wildlife management for the year was $9,870, considerably less than was earned by the Revolving Fund. Villagers started supporting the NPWS wildlife management effort, and local headmen established security committees to prevent poachers from entering their areas.

Once economic benefits started to flow to the local villages, the reduced poaching of elephants has led to an increase of their populations to the level where sustainable harvests would far exceed the total costs of effective management programs. In addition, about half the costs of supporting the village scouts was equivalent to the total sums derived from revenue earned by ivory collected by scouts from elephants which died naturally. While this source of revenue did not go back into the Revolving Fund, it did illustrate to the government the magnitude of funds which could be recovered by this form of local involvement in wildlife management.

In summary, the Wildlife Fund acts as a legal mechanism for charging concession fees, selling wildlife products, and engaging in commercial ventures related to wildlife development. The Fund can then direct the income into appropriate channels to serve the interests of managing the biological resources of the area, as well as the interests of local communities co-existing with the wildlife. It therefore reduces the need to depend on Central Treasury for funds, which in recent years has been unable to meet the growing cost of conservation. A central factor in the success of the Fund was its establishment within NPWD. Using another agency or government body would have diluted the impact of the Fund, particularly its ability to employ local residents as legally-authorized wildlife officers.

(Source: Lewis, Kaweche, and Mwenya, 1987)

CASE STUDY 21: **LAND SWAPS BETWEEN GOVERNMENT AND PRIVATE INDUSTRY: A MECHANISM FOR CONSERVING WETLANDS IN THE USA**

In countries where both government and private industry are major land owners, it is often possible for trades to be arranged that benefit both parties, thereby providing a useful and innovative mechanism for conservation at little or no cost to the taxpayer. One interesting example comes from the USA, where in March 1988 legislation was passed authorizing the exchange of Federal lands in Nevada for privately-owned wetlands in Florida. The Federal Government would then sell the Florida lands to the state and use the proceeds to fund the acquisition of additional lands for two national wildlife refuges in Florida.

Under the exchange agreement, Aerojet-General Corporation will receive title to over 11,000 ha of public lands in Nevada; an additional 5,700 ha of land will be leased to the firm for 99 years. In return, the Federal Government will receive nearly 2,000 ha of wetlands Aerojet owns in south Florida. The Florida land will be sold to the South Florida Water Management District for its use in managing the water resources of southeastern Florida and the Everglades.

The proceeds from that sale, estimated at $2.4 million, then will be used by the US Fish and Wildlife Service to purchase additional lands and inholdings at the Key Deer and Lower Suwannee National Wildlife Refuges in Florida. These purchases will have a significant impact on conservation, providing important habitat for endangered manatees and for wintering waterfowl. The legislation also contains extensive provisions for environmental protection of the Nevada desert transferred to Aerojet, including a specially designated area of over 7,000 ha allocated for the benefit of the desert tortoise.

CASE STUDY 22: **DEBT SWAP FOR CONSERVATION IN ECUADOR: AN INTERNATIONAL INCENTIVE FOR CONSERVING BIOLOGICAL DIVERSITY**

Ecuador is a small South American country with extraordinary levels of biological diversity, containing nearly twice as many species of plants and animals as all of North America. To protect this diversity, 15 protected areas have been established, covering about 11 percent of the land area. As with many Latin American countries, Ecuador is suffering from significant external debt; its 1977 debt balance of $1.3 billion had increased to $9.4 billion by 1987, with 60 percent of the amount owed to private international lenders.

It is apparent to the lenders that Ecuador—like other Latin American countries—is having great difficulties repaying the debt, and the lending banks have recognized this difficulty by reducing the price of Ecuador's debt by 50 percent in the past six months. Further, the debt crisis has generated austerity measures which are seriously hampering development efforts (including sustainable use management of biological resources). After examining the situation, a small group of Ecuadorian professionals, including the former General Managers of Ecuador's Central Bank and of Citibank-Ecuador, organized a private foundation, "Fundación Natura," to use the debt crisis as an opportunity to attract financial resources to be invested in conservation of biological diversity.

Fundación Natura will be in charge of obtaining funds abroad through donations in hard currency. With these funds, a fraction of the Ecuadorean external debt will be purchased at discount value on the secondary financial market. At present, this value fluctuates between 30 and 38 percent of the face value. Fundación Natura will also negotiate donations directly from private lending banks.

The debt notes thus obtained will be exchanged by Fundación Natura for stabilization bonds, thanks to an agreement previously signed between Fundación Natura and the government. The bonds will have the following characteristics:
- *Term:* the same as the purchased external debt (currently nine years);
- *Currency:* sucres;
- *Debt rate of exchange:* Central Bank intervention rate (225 sucres/dollars);

- *Interest rate:* Floating; average of rate paid by the 5 largest Ecuadorian banks on 180-day deposits (30%-40%);
- *Dates of interest payments:* every semester

Fundación Natura will invest the bonds' interest in conservation projects, within the natural areas as defined in the National Conservation Strategy now being prepared. Highest priority will be given to reinforcing National Parks under demographic pressure; conducting scientific research on biological diversity within the protected areas; implementing protection of Galapagos and Machalilla marine reserves; and purchasing natural sanctuaries on the Ecuadorean coastal plains. The amortization of the bonds will be invested in developing Fundación Natura as a professional conservation organization.

In October 1987 the Monetary Board of Ecuador's Central Bank approved the proposal, setting a limit to the swap at 10 million dollars. The first donation, of $1 million dollars from WWF, purchased debt at 35.54% of the face value; each $100 dollars donation purchased $281 of debt. The 180-day deposits currently yield 35%, so a $100 donation will yield $98.35 per year; in nine years the donations from abroad will be multiplied about nine times.

Donations are made to North American NGOs primarily on the basis of existence value, in the hope that donations will provide incentives to developing countries to help them conserve their biological resources. Debt purchase appears to be a very useful way to provide long-term support in local currency, in essence establishing an endowment fund. It is limited by current policies of the US Congress and the Treasury Department, which are currently under review.

Similar debt swaps are being arranged for Costa Rica, Bolivia, the Philippines, and elsewhere, often with support from US-based NGOs such as WWF, Conservation International, and the National Wildlife Federation. The mechanism could also be adapted to debts contracted by Third World governments with multilateral financial institutions such as the World Bank, the International Monetary Fund, the Interamerican Bank; bilateral aid agencies such as USAID, CIDA, SIDA; and debts country-to-country. Debt swaps enable the lender to write off debts if the debtor guarantees to invest the same amount of funds in projects aimed at conserving biological resources.

(Source: Sevilla, 1988)

CASE STUDY 23: **THE INTERNATIONAL TRADE IN TROPICAL TIMBER**

Many tropical countries with large forest resources have provoked wasteful export-oriented "timber booms" by assigning harvesting rights to concessionnaires for royalty, rent, and tax payments that are only a small fraction of the net commercial value of the log harvest (see Case Study 3). They have compounded the damage caused by these incentives by offering only short-term leases, requiring concessionnaires to begin harvesting at once, and adopting royalty systems that encourage loggers to harvest only the best trees while doing enormous damage to the forest land. In response, logging entrepreneurs in several countries have leased virtually the entire productive forest area within a few years and have over-exploited the resource with little concern for future productivity (while unwittingly opening it for clearing by slash-and-burn cultivators).

The result has been wasteful exploitation of the tropical forests, the sacrifice of most of their timber and non-timber values, enormous losses of potential revenue to the government, and at the same time the destruction of rich biological resources. Reforming forest revenue systems and concession terms could raise billions of dollars of additional revenues, promote more efficient, long-term forest resource use, and curtail deforestation.

The promotion of tropical timber imports into certain developed countries, through low tariffs and favorable trade incentives, combined with weak domestic forest policies in the tropical countries and high costs and disincentives to harvesting in the industrial countries, also drives deforestation. The industrial countries typically import unprocessed logs from tropical countries either duty-free or at minimal tariff rates, while imposing much higher tariffs and import restrictions on processed wood products. This encourages developed-country industries to use logs from tropical forests rather than their own, a pattern that is reinforced by domestic restrictions on the amounts that can be cut in domestic forests. The situation may be relieved somewhat by the establishment in 1986 of the International Tropical Timber Organization, based in Yokohama, Japan, which seeks to rationalize trade flows. It is the first commodity agreement that incorporates a specific conservation component.

It is widely recognized that while it is important to monitor the international trade in tropical timber, current practices are inadequate. In particular, there is urgent need for more accurate information on trade in timber species. The International Tropical Timber Agreement aims, among other things, to improve monitoring of the tropical timber trade, both promoting the collection and dissemination of data, and improving international standards and their compatibility.

CASE STUDY 24: **CENTRAL GOVERNMENT SUBSIDIES FOR CONSERVING A WORLD HERITAGE SITE IN QUEENSLAND, AUSTRALIA**

The World Heritage Convention has encouraged the Federal Government of Australia to provide subsidies to States having World Heritage Properties, to compensate for income which might be lost through cessation of extractive uses. In the case of the tropical rainforests of Queensland, the Federal Government has offered the following package to the state, worth some $71.6 million:

- *Public Works:* Up to $13.5 million will be provided for enhancing regional and tourist infrastructure, thereby creating up to 600 permanent jobs by facilitating expansion of tourism. Included are: provision of environmental and recreational parks, interpretive facilities for tourists, rest and parking areas; upgrading of roads and bridges; augmenting water supply and sewerage systems.
- *Reafforestation:* Up to $9.9 million will be spent on reafforestation schemes on both private and public lands, for nursery establishment, planting commercial woodlots on farms, and rehabilitating degraded areas. Up to 270 jobs will be provided.
- *World Heritage Area Management:* Up to $17 million will be allocated for the management, maintenance and presentation of the proposed World Heritage areas, in the provision of tourist and visitor facilities, and the permanent employment of field staff. This will provide 300 jobs.
- *Private Initiatives:* Up to $3.7 million will be available for private initiatives aimed at creating employment for displaced workers and enhancing the attractiveness of the World Heritage area as a tourist destination and otherwise to promote appropriate development in the region. This is expected to create up to 140 job opportunities.
- *Community Initiatives:* Up to $300,000 will be provided for the establishment of a number of community committees to identify regional growth and employment opportunities and allow for these and other proposals to be developed to the feasibility stage.
- *Adjustment Assistance:* Up to $6.5 million has been allocated for a labor adjustment package comprising: the immediate payment of $2,500 dislocation allowance to all eligible

workers; redundancy payments of up to $30,000 for workers 55 and over; wage subsidies (about $100 per week for 26 weeks); training allowance ($30 per week plus unemployment benefit for 13 weeks); and relocation assistance for workers to take up employment in other locations.

- *Business Compensation:* Up to $24.4 million will be provided for businesses directly and substantially affected by the cessation of logging within the World Heritage Area, determined by negotiation. Payments will be made to: major timber companies previously logging inside the World Heritage area; other businesses such as logging contractors, sleeper-cutters, suppliers and rainforest timber users; and a series of payments for the re-tooling of a plywood mill to enable it to process plantation timbers.

CASE STUDY 25: **COOPERATIVE MANAGEMENT OF CARIBOU IN THE ARCTIC**

The Porcupine Caribou (Rangifer tarandus) herd—consisting of over 100,000 adults, plus a similar number of juveniles and calves—has provided meat for food, bones and antlers for tools, and hides for clothing to cultures (especially Inuit and Loucheux) which have occupied northwestern North America for over 20,000 years. Today, many people continue to follow this traditional lifestyle, augmented by modern technology in the form of rifles and snowmobiles. Harvests have ranged between 2,000 and 7,000 adults per year, an offtake which can be sustained by the thriving population.

The Porcupine Caribou Herd is a common property resource shared by the people of the Yukon and the Northwest Territories in Canada and the State of Alaska in the USA. But the proposed exploitation of gas and petroleum in this region, along with supporting infrastructure of roads, towns, and an increased human population, has threatened to disrupt the sustainable utilization of this shared resource. In order to ensure that the relevant parties were able to contribute to improved management of the herd, a Porcupine Caribou Management Agreement was signed in October 1985 by the relevant government agencies and a number of aboriginal people's organizations. It provides a forum for users and managers to discuss caribou issues and to make recommendations on allocation of the available resources.

The Porcupine Caribou Management Board has established a Secretariat, prepared an operating procedure manual, developed a management plan for the herd, and negotiated with US government agencies involved in developing the petroleum resources of the North Slope. The Board's objectives are to cooperatively manage, as a herd, the Porcupine Caribou and its habitat so as to ensure the conservation of the herd, with a view to providing for the ongoing subsistence needs of native users; to provide for the participation of native users in herd management; to recognize and protect certain priority harvesting rights in the herd for native users, while acknowledging that other users may also share the harvest; and to improve communications between governments, native users, and others with regard to management of the herd.

Activities of the Board have included: a biweekly news and information service in English, Loucheux, and Inuktitut about the herd, broadcast by four radio stations which reach every user community; a monthly Porcupine Caribou Almanac in five newspapers; a one-minute television announcement explaining the functions of the Board; guidelines for hunting caribou along the new highway that splits the range of the herd; guidelines on trade and barter of caribou meat; and specifying the Board's position regarding the lease of lands for petroleum exploration and development in the calving grounds included in Alaska's Arctic National Wildlife Refuge.

The Porcupine Caribou Management Board demonstrates how traditional use rights—a major economic incentive, even though it is outside the market system—can be maintained through the use of modern political, communications, and organizational techniques.

Source: PCMB, 1988

DEFINITIONS OF
KEY CONCEPTS

Biological Diversity: The variety and variability among living organisms and the ecological complexes in which they occur (OTA, 1987); often shortened to "biodiversity." "Species diversity" refers to the number of species found within a given area, while "genetic diversity" refers to the variety of genes within a particular species, variety, or breed.

Biological Resources: Living natural resources, including plants, animals, and micro-organisms, plus the environmental resources to which species contribute. Biological resources are the practical target of activities aimed at the principle of conserving biological diversity; they have two important properties the combination of which distinguishes them from non-living resources: they are renewable if conserved; and they are destructible if not conserved (IUCN, 1980).

Buffer Zone: An area on the edge of a protected area which has land use controls which allow only activities compatible with the objectives of the protected area; appropriate activities might include tourism, forestry, agroforestry, etc. The objective of such zones is to give added protection to the reserve, and to compensate local people for the loss of access to the biodiversity resources of the reserve (Oldfield, 1988).

Conservation: The management of human use of the biosphere so that it may yield the greatest sustainable benefit to present generations while maintaining its potential to meet the needs and aspirations of future generations. Thus conservation is positive, embracing preservation, maintenance, sustainable utilization, restoration, and enhancement of the natural environment (IUCN, 1980).

Consumer Surplus: The difference between the total amount of money a consumer would be prepared to pay for some quantity of a good, and the amount he actually has to pay. In economic analysis, consumer surplus is a consideration when the output of the project causes the market price of the product to fall. Those consumers previously paying the higher, old price (what they are willing to pay) will reap a benefit (consumer surplus) from the lower, new price which must be added to the benefits accruing to the new consumers (USAID, 1987).

Cost-Benefit Analysis: The analytical technique used to appraise projects with quantifiable benefits and costs over a finite planning horizon. In project analysis, costs are goods or services used in a project that reduce the benefits of the project; benefits are any goods or services produced by a project that advance the project's objective. In economic analysis, benefits increase the national income of the society while costs reduce the national income of the society. A benefit forgone is a cost just as much as a cost avoided is a benefit. Costs and benefits may be either tangible (land, labor, materials, equipment are tangible costs and increased production of a good or service is a tangible benefit) or intangible (which by definition cannot be directly valued, though they may be quantified in some form).

Debt Swaps: Mechanisms by which part of the external debt of a nation is purchased at a discount and is then sold back to the government in local currency, with the proceeds being used for conservation purposes.

Discount Rate: The interest rate used to determine the present value of a future value by discounting. The opportunity cost of capital is often taken as the discount rate. The "social discount rate," which expresses the preference of a society as a whole for present returns rather than future returns, is used in economic analysis to discount the incremental net benefit stream.

Disincentive: Any inducement or mechanism which discourages governments, local people, and international organizations from depleting biological diversity.

Economic Rent: A value in excess of the costs of production, including a return on the necessary investment. Highly relevant in forestry, where rents collected by concession-holders can be a powerful incentive for increasing production.

Environmental Resource: Resources such as clean air, clean water, scenic values, etc. which are not considered assets; as a result most interest is on activities involved in using these resources and to the ways in which the actions of some users affect the well-being of others.

Externality: A cost which is generated by the producer, but not paid for by him; for example, extracting logs from a hillside may cause increased sedimentation of streams, the cost of which is borne by the downstream farmers instead of the logger. In project analysis, an effect of a project felt outside the project and not included in the valuation of the project. In general, economists consider an externality to exist when production or consumption of a good or service by one economic unit has a direct effect on the welfare of producers or consumers in another unit, without compensation being paid. Detrimental externalities arise if the action is harmful and the agent who carried it out is not charged for the damage done; beneficial externalities arise if the action is beneficial but the agent who carried it out receives no (or insufficient) payment for the benefit. When an externality is quantified in money terms and added to the project accounts, it is said to have been "internalized."

Genetic Resource: A genetic resource is the heritable characteristics of a plant or animal of real or potential benefit to people. The term includes modern cultivars and breeds; traditional cultivars and breeds; special genetic stocks (breeding lines, mutants, etc.); wild relatives of domesticated species; and genetic variants of wild resource species. A "wild genetic resource" is the wild relative of a plant or animal that is already known to be of economic importance. The reasons for conserving such a resource are evident, providing direct and immediate economic benefits; but the genetic material conserved by such a resource must be made available to the people who require it to improve the productivity, quality, or pest resistance of utilized plants or animals.

Incentive (for conserving biological diversity): An incentive is that which incites or motivates desired behavior; for the purposes of these guidelines, an incentive is that which incites or motivates governments, local people, and international organizations to conserve biological diversity. More broadly, an incentive is any inducement on the part of government which attempts to temporarily divert resources such as land, capital, and labor toward conserving biodiversity, and facilitates the participation of certain groups or agents in work which will benefit biodiversity.

Natural Resource: Includes renewable resources (forests, water, wildlife, soils, etc.) and non-renewable resources (oil, coal, iron ore, etc.) which are natural assets.

Opportunity Cost: The benefit forgone by using a scarce resource for one purpose instead of for its best alternative use.

Perverse Incentive: Any incentive which induces behavior leading to the reduction in biological diversity; obviously, "perverse" depends on the perspective, and most perverse incentives are designed to achieve postive policy objectives and the perversity is usually an external factor.

Protected Area: Any area of land which has legal measures which limit human use of the plants and animals within that area; includes national parks, game reserves, multiple-use areas, biosphere reserves, etc.

Shadow Price: The total price or value of an action including, but not limited to, the market price or value. The term is used in economic analysis for a cost or a benefit in a project when the market price is felt to be a poor estimate of economic value.

Subsidy: A subsidy is government economic assistance granted directly or indirectly to individuals or administrative bodies to encourage activities designed to satisfy the needs of the public. It is discretionary and revocable, and is conditional upon certain rules being observed. In contrast to grants, subsidies are usually much more institutionalized and are primarily aimed less at a particular, specific activity than at encouraging works in the public interest.

Sustainable Development: A pattern of social and structural economic transformations (i.e., "development") which optimizes the economic and other societal benefits available in the present, without jeopardizing the likely potential for similar benefits in the future (Goodland and Ledec, 1987).

BIBLIOGRAPHY ON ECONOMIC INCENTIVES AND DISINCENTIVES FOR CONSERVING BIOLOGICAL DIVERSITY

Abelson, P.W. 1979. Property Prices and the Value of Amenities. *Journal of Environmental Economics and Management* 6:11–28.

Adams, Dale W. 1971. Agricultural Credit in Latin America: A Critical Review of External Funding Policy. *American J. Agricultural Economics* 53(2): 163–172.

Ahmad, Yusuf J. 1985. *The Economics of Survival: The Role of Cost-Benefit Analysis in Environmental Decision-Making.* UNEP, Nairobi.

Allegretti, Mary Helena, and S. Schwartzman. 1986. *Extractive Reserves: A Sustainable Development Alternative for Amazonia.* Report to WWF-US, Washington, D.C.

Allen, William H. 1988. Biocultural Restoration of a Tropical Forest. *Bioscience* 38(3):156–161.

Altieri, M.A. and C.L. Merrick. 1987. In Situ Conservation of Crop Genetic Resources Through Maintenance of Traditional Farming Systems. *Econ. Bot.* 41:86–96.

Amacher, R.C., R.D. Tollison and T.D. Willett 1974. The Economics of Fatal Mistakes: Fiscal Mechanisms for Preserving Endangered Predators. *Public Policy* 7:411–441.

Anders, G., W.P. Graham, and S.C. Maurice, 1978. Does Resource Conservation Pay? *International Institute Econ. Research Paper* 14:1–42.

Anderson, C.J. 1973. Animals, Earthquakes and Eruptions. *Field Museum of Natural History Bulletin* 44(5):9–11.

Anderson, Dennis. 1987a. *The Economics of Afforestation: A Case Study in Africa.* Johns Hopkins University Press, Baltimore. 87 pp.

Anderson, Dennis. 1987b. Economic Aspects of Afforestation and Soil Conservation Projects. *Annals of Regional Science* 21(3):100–110.

Anderson, Frederick R. *et al.* 1977. *Environmental Improvement Through Economic Incentives.* Resources for the Future/Johns Hopkins University Press, Baltimore, MD. 195 pp.

Anderson, L.G. 1977. *The Economics of Fisheries Management.* Johns Hopkins University Press, Baltimore, MD.

Archibold, Guillermo. 1988. Kuna Yala Y La Conservacion de Recursos. Paper presented at Workshop on Economics, IUCN General Assembly, 4–5 February 1988, Costa Rica.

Armstrong, F.H. 1968. Valuation of Amenity Forests. *The Consultant* 19:13–19.

Arnold, J.E.M. 1982. *Economic Constraints and Incentives in Agroforestry.* Proceedings of UNU workshop on agroforestry, Freiburg, FRG.

Arrow, K.J. and A.C. Fisher. 1974. Environmental Preservation, Uncertainty, and Irreversibility. *Quarterly J. Economics* 88: 312–319.

Atmosoedarjo, S., L. Daryadi, J. MacKinnon, and P. Hillegers. 1984. National Parks and Rural Communities. In McNeely, J.A. and K.R. Miller (eds.). *National Parks, Conservation and Development: The Role of Protected Areas in Sustaining Society.* Smithsonian Institution Press, Washington, D.C.

Australian Conservation Foundation. 1980. *The Value of National Parks to the Community.* A.C.F. Hawthorn, Victoria. 223 pp.

Ayres, R. and A. Kneese. 1969. Production, Consumption, and Externalities. *American Economic Review* 59(7):

Bachmura, F.T. 1971. The Economics of Vanishing Species. *Natural Resources J.* 11:675–692.

Baker, Alan, Robert Brooks, and Roger Reeves. 1988. Growing for gold. . . and copper. . . and zinc. *New Scientist,* 10 March: 44–48.

Bale, M.D. and E. Lutz. 1981. Price Distortions in Agriculture and Their Effects: An International Comparison. *Am. J. Agricultural Economics* 63(1): 8–22.

Bandyopadhyay, A.K. 1985. The Sunderbans Mangrove Reclamation Experience. *Bakawan* 4(2): 8–10.

Barborak, James R. 1988a. Innovative Funding Mechanisms Used by Costa Rican Conservation Agencies. Paper presented at Workshop on Economics, IUCN General Assembly, 4–5 February 1988, Costa Rica.

Barborak, James R. 1988b. The Role of Economics in Conserving Biological Diversity: A Non-economist's View. Paper presented at Workshop on Economics, IUCN General Assembly, 4–5 February 1988, Costa Rica.

Barclay, A.S. and R.E. Perdue. 1976. Distribution of Anti-Cancer Activity in Higher Plants. *Cancer Treatment Reports* 60(8):1081–1113.

Barnett, Harold J. and C. Morse. 1963. *Scarcity and Growth: The Economics of Natural Resource Availability.* Johns Hopkins University Press, Baltimore, MD.

Barrett, Scott. 1988. Economic Guidelines for the Conservation of Biological Diversity. Paper presented at Workshop on Economics, IUCN General Assembly, 4–5 February 1988, Costa Rica.

Batie, S. and C. Mabbs-Zeno. Opportunity Costs of Preserving Coastal Wetlands: A Case Study of a Recreational Housing Development. *Land Economics* 61(1):1–9.

Baumol, William J. and Wallace E. Oates. 1971. The Use of Standards and Prices to Protect the Environment. *Swedish J. Economics* 73:42–54.

Baumol, William J. and Wallace E. Oates. 1975. *The Theory of Environmental Policy: Externalities, Public Outlays and the Quality of Life.* Prentice-Hall, Englewood Cliffs, New Jersey.

Baxter, W.F. 1974. *People or Penguins: The Case for Optimal Pollution.* Columbia University Press, New York.

Bella, Leslie. 1987. *Parks for Profit.* Harvest House Ltd., Montreal.

Bennett, J. 1984. Using Direct Questioning to Value the Existence Benefits of Preserved National Areas. *Aust. J. Agic. Econ.* 28:136–152.

Bentkover, J., V. Covello, and J. Mumpower. 1986. *Benefits Assessments: The State of the Art.* D. Reidel, Boston.

Berck, R. 1979. Open Access and Extinction. *Econometrica* 47:877–882.

Bingel, A.S. and N.R. Farnsworth. 1980. Botanical Sources of Fertility Regulating Agents: Chemistry and Pharmacology. In *Advances in Hormone Biochemistry and Pharmacology.* Eden Press, New York.

Binswanger, Hans P. 1987. Fiscal and Legal Incentives with Environmental Effects on the Brazilian Amazon. *World Bank Report* ARU 69: 1–48.

Bishop, R.C. 1978. Endangered Species and Uncertainty: the Economics of a Safe Minimum Standard. *Am. J. Agricultural Economics* 60:10–18.

Bishop, R.C. 1979. Endangered Species, Irreversibility, and Uncertainty: A Reply. *Am. J. Agricultural Economics* 61:376–379.

Bishop, R.C. 1980. Endangered Species: An Economic Perspective. *Trans. North Am. Wildl. Nat, Resourc. Conf.* 45:208–218.

Bishop, R.C. 1982. Option value: An Exposition and Extension. *Land Economics* 58(1):1–15.

Bishop, R.C. 1987. Economic Values Defined. In Decker, D.J. and G.R. Goff. *Valuing Wildlife: Economic and Social Perspectives.* Westview Press, San Francisco.

Bockstael, N.E., W.M. Hanemann, and I.E. Strand, Jr. 1984. Measuring the Benefits of Water Quality Improvements Using Recreation Demand Models. U.S. Environmental Protection Agency CR–811043–01–0.

Bohm, P. 1972. Estimating Demand for Public Goods: An Experiment. *European Economic Review* 3:111–130.

Boonkird, S.A. et al. 1984. Forest Villages: An Agroforestry Approach to Rehabilitating Forest Land Degraded by Shifting Cultivation in Thailand. *Agroforestry Systems* 2:87–102.

Borman, F.H. 1976. An Inseparable Linkage: Conservation of Natural Ecosystems and the Conservation of Fossil Energy. *BioScience* 26:754–760.

BOSTID. 1985. Conservation of Biological Diversity in Developing Countries. Report to USAID from the Board on Science and Technology for International Development, National Research Council, Washington D.C.

Boulding, K. 1980. The Economics of the Coming Spaceship Earth. In Daly, H.(ed.), *Economics, Ecology and Ethics.* Freeman, San Francisco.

Bowers, J.K. 1987. The Economics of Wetland Conservation. pp. 157–164 in: Hall, D.O., N. Myers, and N.S. Margaris (eds.). *Economics of Ecosystem Management.* W. Junk Publishers, Dordrecht, Netherlands.

Bradshaw, A.D. and M.J. Chadwick. 1980. The Restoration of Land. Blackwells, Oxford.

Brill, W.J. 1979. Nitrogen Fixation: Basic to Applied. *American Scientist* 67:458–465.

Bristowe, W.S. 1932. Insects and Other Inverterbrates for Human Consumption in Siam. *Trans. Entomological Society London* 80:387–404.

Brooks, K.N., H.N. Gregersen, E.R. Berglund and M. Tayaa. 1982. Economic Evaluation of Watershed Projects: An Overview Methodology and Application. *Water Resource Bulletin* 18(2):245–249.

Brookshire, D.S., A. Randall, and J. Stoll. 1980. Valuing Increments and Decrements in Natural Resources Service Flows. *Am. J. Agric. Econ.* 62:478–488.

Brookshire, D.S., L.S. Eubanks and A. Randall. 1983. Estimating Option Price and Existence Values for Wildlife Resources. *Land Economics* 59(1):1–15.

Browder, John O. 1985. Subsidies, Deforestation and the Forest Sector in the Brazilian Amazon. Report to the World Resources Institute, Washington, D.C.

Browder, John O. 1988. The Social Costs of Rain Forest Destruction: A Critique and Economic Analysis of the "Hamburger Debate." *Interciencia* 13(3):115–120.

Brown, G.M. 1979. Notes on the Economic Value of Genetic Capital: What Can the Blue Whale do for You? Department of Economics, University of Washington, Seattle (MS).

Brown, G.M. 1985. Valuation of Genetic Resources. Paper prepared for Workshop on Conservation of Genetic Resources, Lake Wilderness, WA.

Brown, G.M. and H. Pollakowski. 1977. Economic Value of Shoreline. *Rev. Econ. Stat.* 69:273–278.

Brown, G.M. Jr. and Jon H. Goldstein. 1984. A Model for Valuing Endangered Species. *J. Environmental Economics and Management* 11:303–309.

Brown, P.J. 1984. Benefits of Outdoor Recreation and Some Ideas for Valuing Recreation Opportunities. pp. 209–220 In G.L. Peterson and A. Randall (eds.), *Valuation of Wildland Resource Benefits.* Westview Press, Boulder CO.

Brown, T.C. 1984. The concept of value in resource allocation. *Land Economics* 60:231–246.

Bunker, S.G. 1985. *Underdeveloping the Amazon: Extraction, Unequal Exchange, and the Failure of the Modern State.* University of Illinois Press, Urbana. 279 pp.

Burrows, Paul. 1970. On External Costs and the Visible Arm of the Law. *Oxford Economic Papers* 22:39–56.

Butlin, J.A. (ed.). 1981. *The Economics of Environmental and Natural-Resource Policy.* Westview Press, Boulder, CO.

Buvet, R. et al. (eds). 1977. *Living Systems as Energy Converters.* Elsevier, Amsterdam, Netherlands.

Caldecott, Julian. 1988. *Hunting and Wildlife Management in Sarawak.* IUCN, Gland. 172 pp.

Callicott, J. Baird. 1985. Intrinsic Value, Quantum Theory, and Environmental Ethics. *Environmental Ethics* 7(11):257–275.

Carpenter, R.A. (ed.). 1983. *Natural Systems for Development: What the Planner Needs to Know.* Macmillan, New York.

CEMP (Centre for Environmental Management and Planning). 1988. Preliminary Paper on Compensation for the Establishment of Protected Areas Within Tropical Forest Ecosystems. UNEP, Nairobi.

Cernea, Michael. 1981. Land Tenure Systems and Social Implications of Forestry Development Programmes. World Bank Staff Working Paper 452, World Bank, Washington, D.C.

Chambers, Robert J. 1987. Trees as Savings and Security for the Rural Poor. *IIED Gatekeeper Series* SA3:1–10.

Chang, Sun Joseph. 1982. An Economic Analysis of Forest Taxation's Impact on Optimal Rotation Age. *Land Economics* 58(3): 310–323.

Charbonneau, J.J. and M.J. Hay. 1978. Determinants and Economic Values of Hunting and Fishing. *Trans. N. Am. Wildl. and Natl. Res. Conf.* 43.

Chen, K. 1973. Input-Output Economic Analysis of environmental Impacts. *Transactions on Systems, Man and Cybernetics* 3(6):539–47.

Child, B. and G. Child. 1986. Wildlife, Economic Systems and Sustainable Human Welfare in Semi-arid Rangelands in Southern Africa. *Report on the FAO/Finland Workshop on Watershed Management in Arid and Semi-arid Zones of SADCC Countries,* Maseru, Lesotho. pp. 81–91.

Child, G. 1984. Managing Wildlife for People in Zimbabwe. pp. 118–123 In: McNeely J.A. and Miller K.R. (eds.) *National Parks, Conservation and Development: The Role of Protected Areas in Sustaining Society.* Smithsonian Institution Press, Washington, D.C.

Child, G. 1988a. Economic Incentives and Improved Wildlife Conservation in Zimbabwe. Paper presented at Workshop on Economics, IUCN General Assembly, 4–5 February 1988, Costa Rica.

Child, G. 1988b. Consideration of Institutional Limitations and Reforms for the Better Management of Wildlife in Africa. *Proceedings International Symposium and Conference on Wildlife Utilization in Africa,* Harare, Zimbabwe (in press).

Christiansen, Sofus and Gunnar Poulson (eds.). 1985. *Environmental Aspects of Agricultural Development Assistance.* Nordic Council of Ministers, University of Trondheim, Norway.

Cicchetti, C.J., Joseph H. Seneca, and Paul Davidson. 1969. *The Demand and Supply of Outdoor Recreation: An Econometric Analysis.* Rutgers University Press, New Brunswick, NJ.

Cicchetti, C.J. and M. Freeman. 1971. Option Demand and Consumer Surplus: Further Comment. *Quarterly J. Economics* 85:528–39.

Ciriacy-Wantrup, S.V. 1968. *Resource Conservation: Economics and Policies.* University of California Press, Berkeley. 395 pp.

Ciriacy-Wantrup, S.V. and R. Bishop. 1975. Common Property as a Concept in Natural Resources Policy. *Natural Resources* 15:713–27.

Ciriacy-Wantrup, S.V. and W.E. Phillips 1970. Conservation of the California Tule Elk: A Socioeconomic Study of a Survival Problem. *Biological Conservation* 3(1):23–32.

Clark, Colin W. 1973a. The Economics of Overexploitation. *Science* 181:630–634.

Clark, Colin W. 1973b. Profit Maximization and the Extinction of Animal Species. *J. Pol. Econ.* 81:950–961.

Clark, Colin W. 1976. *Mathematical Bioeconomics: The Optimal Management of Renewable Resources.* John Wiley, New York.

Clark, E.H. 1971. Multipart Pricing of Public Goods. *Public Choice* 11(3):17–33.

Clark, Vance L. 1988. Implementing Conservation Easements. *J. Soil and Water Conservation* Jan–Feb:31–32.

Clawson, Marion and Jack Knetsch. 1966. *Economics of Outdoor Recreation.* Johns Hopkins University Press, Baltimore, MD.

Cocheba, Donald J., and William A. Langford. 1978. Wildlife Valuation: the Collective Good Aspect of Hunting. *J. Land Econ.* 54(4):490–504.

Connaughton, C.A. 1943. Yield of Water as an Element in Multiple Use of Wildland. *J. Forestry* 14:641–644.

Conrad, Jon M. and Colin W. Clark. 1987. *Natural Resource Economics.* Cambridge University Press, Cambridge, UK.

Coomber, N.H. 1973. *Evaluation of Environmental Intangibles.* Genera Press, New York.

Cooper, C. 1981. *Economic Evaluation and the Environment.* Hodder and Stoughton, London.

Cropper, M.L., Milton C. Veinstein and Richard J. Zeckhauser. 1978. The Optimal Consumption of Depletable Natural Resources: An Elaboration, Correction, and Extension. *Quarterly J. Economics* 92:337–353.

Crutchfield, J.A. 1962. Valuation of Fishery Resources. *Land Econ.* 38(5):145–154.

Cuddington, J.T., F.R. Johnson, and J.L. Knetsch, 1981. Valuing Amenity Resources in the Presence of Substitutes. *Land Economics* 57:526–535.

Cumming, D.H.M. 1985. Environmental Limits and Sustainable Harvests. Paper presented to the Plenary Session of the Zimbabwe National Conservation Strategy Conference, Harare.

Cummings, R., D. Brookshire, and W. Schulze. 1986. *Valuing Environmental Goods: A State of The Art Assessment of the Contingent Valuation Method.* Rowman and Allenheld, Totowa, NJ. 270 pp.

Dalfelt, A. 1976. Some Data Related to Costs and Benefits of National Parks in Latin America. Mimeo report, CATIE, Costa Rica. 66 pp.

Daly, H.E. 1980. *Economics, Ecology and Ethics.* Freeman, San Francisco, CA.

Daly, H.E. and Umana, A.F. (Eds). 1981. *Energy, Economics, and the Environment.* Westview Press, Boulder, CO.

Daniel, J.G. and A. Kulasingham. 1974. Problems Arising from Large-scale Forest Clearing for Agricultural Use. *Malaysian Forester* 37:152–160.

D'Arge, R., R. Ayres, and A. Kneese. 1972. *Economics and Environment: A Materials Balance Approach.* Johns Hopkins Press, Baltimore, MD.

Dasgupta, P.S. and G.M. Heal. 1979. *Economic Theory and Exhaustible Resources.* Cambridge University Press, Cambridge, UK.

Dasgupta, P. 1982. *The Control of Resources.* Harvard University Press, Cambridge, MA.

Dasmann, R.F., Milton, J.P. and P. Freeman. 1973. *Ecological Principles for Economic Development.* John Wiley, London. 252 pp.

Davis, R.K. 1964. The Value of Big Game Hunting in a Private Forest. *Transactions Twenty-Ninth North American Wildlife and Natural Resources Conference.*

Decker, D. and G. Goff, 1987. *Valuing Wildlife: Economic and Social Perspectives.* Westview Press, Boulder, CO.

de Groot, R.S. 1986. *A Functional Ecosystem Evaluation Method as a Tool in Environmental Planning and Decision Making.* Agricultural University, Wageningen, The Netherlands. 38 pp.

de Groot, R.S. 1987. Environmental Functions as a Unifying Concept for Ecology and Economics. *Environmentalist* 7(2):105-109.

de Groot, R.S. 1988. The Use of Economic Incentives to Promote Biological Diversity: A Case Study of the Galapagos Islands. Paper presented at Workshop on Economics, IUCN General Assembly, 4-5 February 1988, Costa Rica.

de Koning, H.W. 1987. *Setting Environmental Standards: Guidelines for Decision-Making.* WHO, Geneva. 98 pp.

Delogu, Orlando E. and Hermann Soell. 1976. Fiscal Measures for Environmental Protection: Two Divergent Views. *IUCN Environmental Policy and Law Paper* 11:1-79.

Demsetz, H. 1964. The Exchange and Enforcement of Property Rights. *J. Law and Economics* 7:11-26.

Dennis, E. 1981. A Method for Designing Cost-effective Wilderness Allocation Alternatives. *Forest Science* 27(3): 551-566.

Dickenson, R.E. 1981. Effects of Tropical Deforestation on Climate. In: *Blowing in the Wind: Deforestation and Long-range Implications.* Studies in Third World Societies 14. College of William and Mary, Williamsburg, VA.

Diegues, Antonio Carlos. 1987. Biological Diversity, Economic Incentives, and Traditional Coastal Cultures in Brazil. Paper presented at Workshop on Economics, IUCN General Assembly, 4-5 February 1988, Costa Rica.

Dixon, John A. 1987. Managing Watershed Resouces. *Annals of Regional Science* 21(3):111-123.

Dixon, John A., and M. Hufschmidt. 1986. *Economic Valuation Techniques for the Environment—A Case Study Workbook.* Johns Hopkins University Press, Baltimore.

Dixon, John A. *et al.* 1986. *Economic Analysis of the Environmental Impacts of Development Projects.* Asian Development Bank, Manila. 100 pp.

Dixon, John A. and S. Wattanavitukul. 1982. *A Summary of Work on Benefit-Cost Analysis of Natural Systems and Environmental Quality Aspects of Development.* Environment and Policy Institute, East-West Center, Honolulu.

Dorfman, R. and Nancy S. Dorfman (eds.). 1977. *Economics of the Environment.* W.W. Norton and Co., New York. 510 pp.

Downing, Paul B. 1984. *Environmental Economics and Policy.* Little, Brown and Co., Boston, MA.

Dumsday, R.G., D.A. Oram, and S.E. Lumley. 1983. Economic Aspects of the Control of Dryland Salinity. *Proc. Royal Society of Victoria* 95(3):139–145.

Dwyer, John, John Kelly, and Michael Bowes. 1977. *Improved Procedures for Valuation of the Contribution of Recreation to National Economic Development.* Report No. 128. University of Illinois, Water Resources Centre, Urbana-Champaign. 218 pp.

Edwards, Steven. 1987. In Defense of Environmental Economics. *Environmental Ethics* 9(1):74–85.

Ehrenfeld, D.W. 1976. The Conservation of Non-resources. *Am. Sci.* 64:648–56.

Ehrenfeld, David. 1988. Why Put a Value on Biodiversity? pp. 212–216. In Wilson, E.O. (ed.). *Biodiversity.* National Academy Press, Washington D.C.

Ehrlich, P.R. and A.H. Ehrlich. 1981. *Extinction: The Causes and Consequences of the Disappearance of Species.* Random House, New York. 305 pp.

Eltringham, S.K. 1984. *Wildlife Resources and Economic Development.* John Wiley, New York. 325 pp.

Esterlin, Richard. 1973. Does Money Buy Happiness? *Public Interest* 30(1):3–10.

Everett, R.D. 1978. The Monetary Value of the Recreational Benefits of Wildlife. *J. Environmental Management* 8:203–13.

Farnsworth, N.R. 1982. The Potential Consequences of Plant Extinction in the United States on the Current and Future Availability of Prescription Drugs. Paper presented at Symposium on Estimating the Value of Endangered Species—Responsibilities and Role of the Scientific Community, Annual Meeting of the AAAS, Washington, D.C.

Farnsworth, N.R. and R.W. Morris. 1976. Higher Plants: The Sleeping Giant of Drug Development. *American J. Pharmacy* 148(2):46–52.

Farnsworth, N.R. and D.D. Soejarto. 1985. Potential Consequences of Plant Extinction in the United States on the Current and Future Availability of Prescription Drugs. *Economic Botany* 39(2):231–240.

Fearnside, P.M. 1980. The Effects of Cattle Pasture on Soil Fertility in the Brazilian Amazon: Consequences for Beef Production Sustainability. *Tropical Ecology* 21(1): 125–137.

Feit, Harvey A. 1988. Self-Management and State-Management: Forms of Knowing, and Conserving. Paper presented at Smithsonian Institution Symposium on Culture and Conservation, April 8–9.

Fillon, F.L., A. Jacquemot, and R. Reid. 1985. The Importance of Wildlife to Canadians. Canadian Wildlife Service, Ottawa. 19 pp.

Finney, C.E. and S. Western. 1986. An Economic Analysis of Environmental Protection and Management: An Example from the Philippines. *Environmentalist* 6(1):45–61.

Fisher, A.C. 1973. Environmental Externalities and the Arrow-Lind Public Investment Theorem. *Am. Econ. Rev.* 63(4):722–725.

Fisher, A.C. 1981a. Economic Analysis and the Extinction of Species. *Report No. ERG–WP–81–4.* Energy and Resources Group, Berkeley, CA. 19 pp.

Fisher, A.C. 1981b. *Resources and Environmental Economics.* Cambridge University Press, Cambridge, UK. 284 pp.

Fisher, A.C and M. Hanemann. 1984. Option Values and the Extinction of Species. *Working Paper No 269. Giannini Foundation of Agricultural Economics,* Berkeley, CA. 39pp.

Fisher, A.C. and W.M. Hanemann, 1985. Endangered Species: the Economics of Irreversible Damage. pp 129–138 In D.O. Hall, N. Myers and N.S. Margaris (eds.), *Economics of Ecosystem Management.* W. Junk Publishers, Dordrecht, The Netherlands.

Fisher, A.C. and J.V. Krutilla. 1985. Resource Conservation, Environmental Preservation, and the Rate of Discount. *Quarterly J. Economics* 89:358–70.

Fisher, A.C., J.V. Krutilla, and C.J. Cicchetti, 1972. The Economics of Environmental Preservation: A Theoretical and Empirical Analysis. *Am. Econ. Rev.* 62:605–619.

Fisher, A.C. and F.M. Peterson. 1976. The Environment in Economics: A Survey. *Journal of Economic Literature* 14:1–33.

Fitter, Richard. 1986. *Wildlife for Man: How and Why We Should Conserve Our Species.* Collins, London. 223 pp.

Forsund, F. 1985. Input-Output Models, National Economic Models, and the Environment. In Kneese, A.V. and J.L. Sweeney (eds.). *Handbook of Natural Resource and Energy Economics.* Elsevier, New York.

Fortmann, Louise. 1985. The Tree Tenure Factor in Agro-forestry with Particular Reference to Africa. *Agro-forestry Systems* 2:229–251.

Fox, A.M. 1984. People and Their Park. An Example of Free Running Socio-Ecological Succession. In: McNeely, J.A. and K.R. Miller (eds.). *National Park, Conservation, and Developmnet: The Role of Protected Areas in Sustaining Society.* Smithsonian Institution Press, Washington, D.C.

Frankel, O.H. and J.G. Hawkes (eds). 1974. *Plant Genetic Resources for Today and Tomorrow.* Cambridge University Press, London, UK.

Frankel, O.M. and Michael E. Soulé. 1981. *Conservation and Evolution.* Cambridge University Press, New York. 327 pp.

Freeman, A.M. III 1979. *The Benefits of Environmental Improvement: Theory and Practice.* Johns Hopkins University Press, Baltimore.

French, James H. and Romeo H. Gecolea. 1986. *A Forester's Guide for Community Involvement in Upland Conservation.* FAO, Rome. 125 pp.

Ferguson, I.S. and P.J. Grieg. 1973. What Price Recreation? *Australian Forestry* 36:80–90.

Garcia, Jose Rafael. 1984. Waterfalls, Hydropower, and Water for Industry: Contributions from Canaima National Park. pp. 588–591 In McNeely, J.A. and K.R. Miller (eds.), *National Parks, Conservation, and Development: The Role of Protected Areas in Sustaining Society.* Smithsonian Institution Press, Washington, D.C.

Garratt, Keith. 1984. The Relationship Between Adjacent Lands and Protected Areas: Issues of Concern for the Protected Area Manager. pp 65–71 in McNeely, J.A. and K.R. Miller (eds.). *National Parks, Conservation, and Development: The Role of Protected Areas in Sustaining Society.* Smithsonian Institution Press, Washington, D.C.

Garrison, C.B. 1974. A Case Study of the Local Economic Impact of Reservoir Recreation. *J. Leisure Research* 6:7–19.

Gibson, J.G. and R.W. Anderson. 1975. The Estimation of Consumers' Surplus from a Recreation Facility with Optional Tariffs. *Applied Economics* 7:73–79.

Gilbert, A.J. and W.A. Hafkemp. 1986. Natural Resource Accounting in a Multi-Objective Context. *Annals of Regional Science* 20(3):10–37.

Giles, Robert H. 1971. *Wildlife Management Techniques* (3rd edition). The Wildlife Society, Washington, D.C. 633 pp.

Gillis, M. 1986. *Non-wood Forest Products in Indonesia.* Department of Forestry, University of North Carolina, Chapel Hill, North Carolina.

Goodin, R. 1982. Discounting Discounting. *J. Public Policy* 2:

Goodland, R. 1984. The World Bank, Environment, and Protected Areas. In: McNeely, J.A. and K.R. Miller (eds.). *National Parks, Conservation and Development: The Role of Protected Areas in Sustaining Society.* Smithsonian Institution Press, Washington, D.C.

Goodland, R. 1988. A Major New Opportunity to Finance Biodiversity Preservation. pp. 437–445 in Wilson, E.O. *Biodiversity.* National Academy Press, Washington D.C.

Gordon, H.S. 1954. The Economic Theory of a Common Property Resource: The Fisher. *J. Pol. Econ.* 124–142.

Gordon, I.M. and J.L. Knetsch. 1979. Consumer's Surplus Measures and the Evaluation of Resources. *Land Econ.* 55:1–27

Gosselink, J.G., E.P. Odum, and R.M. Pope. 1973. The Value of the Tidal Marsh. Paper LSU–SG–74–03. Center for Wetland Resources, Louisiana State Univ. Baton Rouge, LA. 23pp.

Goulding, M. 1980. *The Fishes and the Forest: Explorations in Amazonian Natural History.* University of California Press, Berkeley. 280 pp.

Government of India. 1983. *Eliciting Public Support for Wildlife Conservation.* Report of Indian Board for Wildlife Task Force, Department of Environment, New Delhi.

Graham, Daniel A. 1981. Cost Benefit Analysis Under Uncertainty. *American Economic Review* 71:15–725.

Gray, John W. 1983. *Forestry Revenue Systems in Developing Countries.* FAO Forestry Paper 43, Rome.

Greenley, D.A., R.G. Walsh and R.A. Young. 1981. Option Value: Empirical Evidence from a Case of Recreation and Water Quality. *Quarterly J. Economics* 95:657–673.

Gregersen, Hans. 1983. Incentives for Afforestation: A Comparative Assessment. Presented at the International Symposium on Strategies and Designs for Afforestation, September 19–23, Waggeningen, The Netherlands.

Gregersen, Hans, T. Houghtaling and A. Rebenstein. 1979. Economics of Public Forestry Incentive Programs: A Case Study of Cost-Sharing in Minnesota. Agricultural Experiment Station, Technical Bulletin No. 315. University of Minnesota.

Gregersen, H.M. and S.E. McGaughy. 1985. *Improving Policies and Financing Mechanisms for Forestry Development.* Inter-American Development Bank, Washington D.C. 110 pp.

Gregersen, H.M. et al. 1987. Guidelines for Economic Appraisal of Watershed Management Projects. *FAO Conservation Guide* 16:1–144.

Gregory, David D. 1972. The Easement as a Conservation Technique. *IUCN Environmental Law Paper* 1:1–47.

Gregory, G.R. 1955. An Economic Approach to Multiple Use. *Forest Science* 1:13–20.

Grojean, R.E., J.A. Sousa, and M.C. Henry. 1980. Utilization of Solar Radiation by Polar Animals: An Optical Model for Pelts. *Applied Optics* 19(3):339–346.

Grow, G.S. 1987. Polar Bears Have Solar Hairs. *Christian Science Monitor* November 2–8.

Gupta, T. and A. Guleria. 1982. *Non-wood Forest Products from India.* IBH Publishing Co., New Delhi.

Gupta, T.A. and J.H. Foster. 1975. Economic Criteria for Freshwater Wetland Policy in Massachusetts. *American J. Agricultural Economics* 57(1):40–45.

Hahn, C. 1982. *The Economic Rationale for Protection and Management of Natural Areas in Developing Countries.* Natural Resources Defence Council, Washington, D.C. 47 pp.

Hair, Jay D. 1988. *The Economics of Conserving Wetlands: A Widening Circle.* Paper presented at Workshop on Economics, IUCN General Assembly, 4–5 February 1988, Costa Rica.

Hall, Charles A., C.J. Cleveland, and Robert Kaufmann. 1986. *Energy and Resource Quality: The Ecology of the Economic Process.* John Wiley and Sons, New York. 577 pp.

Hall, D.O., N. Myers, and N.S. Margaris (eds.). 1985. *Economics of Ecosystem Management.* W. Junk Publishers, Dordrecht, The Netherlands.

Hamilton, Lawrence S. and Jeff M. Fox. 1987. Protected Area Systems and Local People. Paper presented at Workshop on Fields and Forests, Xishuangbanna, Yunnan, China.

Hammack, J. and G.M. Brown, Jr. 1974. *Waterfowl and Wetlands: Toward Bioeconomic Analysis.* Johns Hopkins University Press, Baltimore, MD.

Hanemann, W. Michael. 1988. Economics and the Preservation of Biodiversity. pp. 193–199 In Wilson, E.O. (ed.). *Biodiversity.* National Academy Press, Washington, D.C.

Harris, Stuart. 1985. The Economics of Ecology and the Ecology of Economics. *Search* 16(9–12):284–290.

Harris, S., and A. Ulph. 1975. The Economics of Environmental Services. In K. Tucker (ed.). *The Economics of the Service Sector.*

Hartwick, J.M. and N.D. Olewiler. 1986. *The Economics of Natural Resource Use.* Harper and Row, New York.

Helfand, Gloria E. 1986. *Timber Economics and Other Resource Values.* The Wilderness Society, Washington, D.C.

Helliwell, D.R. 1969. Valuation of Wildlife Resources. *Regional Studies* 3:41–47.

Helliwell, D.R. 1973. Priorities and Values in Nature Conservation. *J. Environmental Management* 1:85–127.

Helliwell, D.R. 1975. Discount Rates and Environmental Conservation. *Environmental Conservation* 2(2):199–201.

Helliwell, D.R. 1978. Preparation of Value Indices for Non-market Benefits Based on Characteristics of the Forest Environment. Paper presented at World Forestry Congress, Jakarta.

Henderson-Sellers, A. 1981. The Effects of Land Clearance and Agricultural Practices on Climate. pp 443–486 In *Blowing in the Wind: Deforestation and Long-range Implications.* Studies in Third World Societies 14, College of William and Mary, Williamsburg, VA.

Henry, C. 1974. Option Values in the Economics of Irreplaceable Assets. *Review of Economic Studies* 89–104.

Herfindahl, Orris C., and Allen Kneese. 1974. *Economic Theory of Natural Resources.* Merrill, Columbus, OH.

Heyman, Arthur M. 1988. Self-Financed Resource Management: A Direct Approach to Maintaining Marine Biological Diversity. Paper presented at Workshop on Economics, IUCN General Assembly, 4–5 February 1988, Costa Rica.

Hirsch, Fred. 1976. *Social Limits to Growth.* Harvard University Press, Cambridge, MA.

Hirshleifer, J. 1977. Economics from a Biological Viewpoint. *J. Law and Economics.* 30:

Howe, Charles W. 1979. *Natural Resource Economics: Issues, Analysis, and Policy.* John Wiley and Sons, New York.

Hueting, R. 1985. Results of an Economic Scenario That Gives Top Priority to Saving the Environment Instead of Encouraging Production Growth. *Environmentalist* 5(4):253–262.

Hufschmidt, Maynard M. et al. 1983. *Environment, Natural Systems, and Development: An Economic Valuation Guide.* Johns Hopkins University Press, Baltimore, MD. 338 pp.

Hufschmidt, Maynard M. and Ruangdej Srivardhana. 1986. The Nam Pong Water Resources Project in Thailand. pp. 141–162 in Dixon, John A., and M. Hufschmidt (eds.). *Economic Valuation Techniques for the Environment—A Case Study Workbook.* Johns Hopkins University Press, Baltimore, MD.

Hyde, William F. 1983. Development Versus Preservation in Public Resource Management: A Case Study from the Timber-Wilderness Controversy. *J. Environmental Management* 16:347–355.

Island Resources Foundation. 1981. Economic Impact Analysis for the Virgin Islands National Park. US Dept. of the Interior, National Park Service, Washington, D.C.

Israel, A. 1987. *Institutional Development: Incentives for Performance.* Johns Hopkins University Press, Baltimore, MD.

IUCN. 1980. *World Conservation Strategy: Living Resource Conservation for Sustainable Development.* IUCN–UNEP–WWF, Gland. 44 pp.

IUCN. 1985. *1985 United Nations List of National Parks and Protected Areas.* IUCN, Gland, Switzerland. 174 pp.

IUCN. 1986a. *Review of the Protected Areas System in Oceania.* IUCN, Gland, Switzerland. 239 pp.

IUCN. 1986b. *Review of the Protected Areas System in the Afrotropical Realm.* IUCN, Gland, Switzerland. 259 pp.

Izac, A.-M.N. 1986. Resources Policies, Property Rights and Conflicts of Interest. *Australian J. Agricultural Economics* 30(1):23–27.

Jacobs, P. and D. Munro (eds.). *Conservation With Equity: Strategies for Sustainable Development.* IUCN, Gland. 466 pp.

James, D.E., H.M.A. Jansen, and J.B. Opschoor. 1978. *Economic Approaches to Environmental Problems.* Elsevier, Amsterdam, The Netherlands.

Janzen, Dan. 1988. The Use of Economic Incentives in Costa Rica's Guanacaste National Park. Paper presented at Workshop on Economics, IUCN General Assembly, 4–5 February 1988, Costa Rica.

Jefferies, Margaret. 1985. *The Story of Mt. Everest National Park.* Cobb/Hopwood Publications, Auckland, New Zealand. 192 pp.

Johansson, P-O. 1987. *The Economic Theory and Measurement of Environmental Benefits.* Cambridge University Press, London. 238 pp.

Johnson, M. and J. Bennet. 1981. Regional Environmental Science and Economic Impact Evaluation. *Regional and Urban Economics* 11(2):215–230.

Johnson, R.L. and G.V. Johnson (eds.). 1984. *Economic Valuation of Natural Resources: Issue, Theory and Applications.* Westview Press, Boulder, CO.

Jones, J. Greg, W.G. Beardsley, D.W. Countryman, and Dennis L. Schweitzer. 1978. Estimating economic costs of allocating land to wilderness. *Forest Sci.* 24(3):410–422.

Jones-Lee, M. W. 1976. *The Value of Life: An Economic Analysis.* University of Chicago Press, Chicago.

Kalter, R.J. and W.B. Lord. 1968. Measurement of the Impact of Recreation Investments on a Local Economy. *American J. Agricultural Economics* 50:243–256.

Kelman, Steven. 1981. *What Price Incentives: Economics and the Environment.* Auburn House, Boston MA.

Kennedy, Duncan. 1980. Cost-Benefit Analysis of Entitlement Problems: A Critique. *Standard Law Review* 33(2):419–431.

Klee, G.A. (ed). 1980. *World Systems of Traditional Resource Management.* John Wiley and Sons, New York. 290 pp.

Kneese, A.V. 1971. Environmental Economics and Policy. *The American Economic Review* 61:153–166.

Kneese, A.V., R.V. Ayres, and R. D'Arge. 1970. *Economics and the Environment: A Materials Balance Approach.* Johns Hopkins University Press, Baltimore, MD.

Kneese, A.V. and J.L. Sweeney, (Eds). 1985. *Handbook of Natural Resources and Energy Economics.* Elsevier, New York.

Krutilla, J.V. 1967. Conservation Reconsidered. *American Economic Review* 57(4):

Krutilla, J.V. and C.J. Cicchetti. 1972. Evaluating Benefits of Environmental Resources, with Special Application to the Hells Canyon. *Natural Resources J.* 12:1–29.

Krutilla, J.V. and A.C. Fisher. 1975. *The Economics of Natural Environments: Studies in the Valuation of Commodity and Amenities Resources.* Resources for the Future/Johns Hopkins University Press, Baltimore, MD. 292 pp.

Krutilla, J.V. and A.C. Fisher. 1980. Valuing Long-run Ecological Consequences and Irreversibilities. *J. Environmental Economics and Management* 1(2):96–108.

Kumazaki, Minoru. 1982. Sharing Financial Responsibility with Water Users for Improvement of Forested Watersheds. A Historical Review of the Japanese Experience. In R. Handa (ed.). *The Current State of Japanese Forestry.* Contribution to the XVII IUFRO Congress, Kyoto, Japan.

Kux, Molly. 1986. Land Use Options to Conserve Living Resources and Biological Diversity in Developing Countries. Manuscript, University of Florida.

Kwapena, N. 1984. Wildlife Management by the People. In McNeely, J.A. and K.R. Miller (eds.). *National Parks, Conservation, and Development: The Role of Protected Areas in Sustaining Society.* Smithsonian Institution Press, Washington, D.C.

Langford, William A. and Donald J. Cocheba. 1978. The Wildlife Valuation Problem: a Critical Review of Economic Approaches. *Can. Wildl. Serv. Occ. Paper* 37: 1–35.

Lausche, B.J. 1980. Guidelines for Protected Area Legislation. IUCN Environmental Policy and Law Paper 16: 1–108.

Lecomber, J.R.C. 1979. *The Economics of Natural Resources.* Macmillan, London.

Ledec, G. 1985. The Political Economy of Tropical Deforestation. pp. 179–226 in Leonard, H.J. (ed.). *Divesting Nature's Capital: The Political Economy of Environmental Abuse in the Third World.* Holmes and Meier, New York. 299 pp.

Ledec, G. and R. Goodland. 1986. Epilogue. in Schumann, D.A. and W.L. Partridge (eds.). *The Human Ecology of Tropical Land Settlement in Latin America.* Westview Press, Boulder, CO.

Lee, K.S. 1982. A Generalised Input-Output Model of an Economy with Environmental Protection. *Review of Economics and Statistics* 65(3):466–473.

Leipert, Christian. 1986. Social Costs of Economic Growth. *J. Economic Issues* 20(1):109–131.

Leontief, W. 1970. Environmental Repercussions and Economic Structure: An Input-Output Approach. *Review of Economics and Statistics* 52(3):262–271.

Leslie, A.J. 1987. A Second Look at Timber Economics of Natural Management Systems in Tropical Mixed Forests. *Unasylva* 59(1):46–58.

Levin, D.A. 1976. The Chemical Defenses of Plants to Pathogens and Herbivores. *Annual Review of Ecology and Systematics* 7:121–159.

Lewis, Dale M., G.B. Kaweche, and Ackim Mwenya. 1987. Wildlife Conservation Outside Protected Areas: Lessons from an Experiment in Zambia. *Lupande Research Project Publication* 4:1–14.

Lewis, Harrison F. 1951. Wildlife in Today's Economy: Aesthetic and Recreational Values of Wildlife. *Trans. N. Am. Wildl. Conf.* 16:13–16.

Livingstone, Ian. 1986. The Common Property Problem and Pastoralist Economic Behaviour. *J. Development Studies* 23:5–19.

Loomis, J.B. 1986. Assessing Wildlife and Environmental Values in Cost-Benefit Analysis: State of the Art. *J. Environmental Management* 22:125–131.

Lothian, Andrew. 1985. A Cost-benefit Study of National Parks on Kangaroo Island, South Australia. In *Proceedings of the Conference on Conservation and the Economy 1984.* Australian Government Publishing Service, Canberra.

Lucas, P.H.C. 1984. How Protected Areas Can Help Meet Society's Evolving Needs. In McNeely, J.A. and K.R. Miller (eds.). *National Parks, Conservation, and Development: The Role of Protected Areas in Sustaining Society.* Smithsonian Institution Press, Washington, D.C.

Lusigi, Walter J. 1978. *Planning Human Activities on Protected Natural Ecosystems.* Dissertationes Botanicae 48. J. Cramer, Vaduz, Germany. 233 pp.

Lusigi, Walter. 1984. Mt. Kulal Biosphere Reserve: Reconciling Conservation with Local Human Population Needs. pp. 459–469 in McNeely, J.A. and D. Navid. *Conservation, Science, and Society.* Unesco-UNEP, Paris.

Lynne, G., P. Conroy, and F. Prochaska. 1981. *Economic Valuation of Marsh Areas for Marine Production. J. Environmental Economics and Management* 8:175–186.

Lyster, S. 1985. International Wildlife Law. *IUCN Environmental Policy and Law Paper* 22:1–470.

MacKinnon, J.R. 1983. Irrigation and Watershed Protection in Indonesia. Report to the World Bank.

MacKinnon, J.R., K. MacKinnon, G. Child, and J. Thorsell. 1986. *Managing Protected Areas in the Tropics.* IUCN, Gland. 295 pp.

MacKinnon, J.R. and K. MacKinnon. 1986. *Review of the Protected Areas System in the Indo-Malayan Realm.* IUCN, Gland. 284 pp.

Macleod, Scott. 1974. Financing Environmental Measures in Developing Countries: The Principle of Additionality. *IUCN Environmental Policy and Law Paper* 6:1–54.

Maler, K.G. 1977. A Note on the Use of Property Values in Estimating Marginal Willingness to Pay for Environmental Quality. *J. Environmental Economics and Management* 4:355–69.

Marglin, S. 1963. The Social Rate of Discount and the Optimal Rate of Investment. *Q.J. Econ.* 77(1):95–111.

Markandya, A. and D. Pearce. 1987. Natural Environments and the Social Rate of Discount. Discussion Paper No. 27, Department of Economics, University College London.

Martin, R. 1986. Communal Areas Management Programme for Indigenous Resources (CAMPFIRE). Branch of Terrestrial Ecology, Working Document No. 1/86, Department of National Parks and Wildife Management. 110 pp.

Martin, R.B. and V. Clarke. 1988. Predicted Returns for Wildlife Management in the Omay Communal Land. Annex to the Land Use Study in Omay Communal Land, Zimbabwe, Agricultural and Rural Development Authority, Harare.

McConnell, K.E. and J.G. Sutinen. 1979. Bioeconomic models of marine recreational fishing. *J. Env. Econ. and Mgt.* 6:127–139.

McMichael, D.F. (ed.). 1971. *Society's Demand for Open Air Recreation, Wilderness, and Scientific Reference Areas.* Institute of Australian Foresters, Canberra.

McNeely, J.A. 1987. How Dams and Wildlife Can Co-Exist: Natural Habitats, Agriculture, and Major Water Resource Development Projects in Tropical Asia. *J. Conservation Biology* 1(3): 228–238.

McNeely, J. A. and K.R. Miller (eds.). 1984. *National Parks, Conservation, and Development: The Role of Protected Areas in Sustaining Society.* Smithsonian Institution Press, Washington, D.C. 838 pp.

McNeely, J. A., Kenton R. Miller, and James W. Thorsell. 1987. Objectives, Selection, and Management of Protected Areas in Tropical Forest Habitats. pp. 181–204 in *Primate Conservation in the Tropical Rain Forest.* Alan R. Liss, Inc., New York.

McNeely, J.A. and D. Navid (eds.). 1984. *Conservation, Science and Society: The Proceedings of the First International Congress on Biosphere Reserves.* Minsk, Byelorussia, USSR. 600 pp.

McNeely, J.A. and David Pitt (eds.). 1984. *Culture and Conservation: The Human Dimension in Environmental Planning.* Croom Helm, London. 308 pp.

McNeely, J.A. and J.W. Thorsell (eds.). 1985. *People and Protected Areas in the Hindukush-Himalaya.* ICIMOD, Kathmandu. 250 pp.

McNeely, J.A. and J.W. Thorsell. 1987. Guidelines for Development of Terrestrial and Marine National Parks for Tourism and Travel. World Tourism Organization, Madrid. 29 pp.

Meeks, Gordon, Jr. 1982. State Incentives for Non-Industrial Private Forestry. *J. Forestry* 82(1):18–22.

Merriam, Larry C. 1964. The Bob Marshall Wilderness Areas of Montana: Some Socioeconomic Considerations. *J. Forestry* 62(11): 789–795.

Messerschmidt, Don. 1985. People's Participation in Park Resource Planning and Management. pp. 133–140 in McNeely, J.A. and J.W. Thorsell (eds.). 1985. *People and Protected Areas in the Hindukush-Himalaya.* ICIMOD, Kathmandu. 250 pp.

Miller, David L. 1986. Technology, Territoriality and Ecology: The Evolution of Mexico's Caribbean Spiny Lobster Fishery. Paper presented at Workshop on Ecological Management of Common Property Resources, IV International Congress of Ecology, Syracuse, New York.

Miller, J.R. 1981. Irreversible Land Use and the Preservation of Endangered Species. *J. Environmental Economics and Management* 8:19–26.

Miller, J.R. and F.C. Menz 1979. Some Economic Considerations for Wildlife Preservation. *Southern Economic J.* 45(3):718–729.

Mishan, E.J. 1971. The Postwar Literature on Externalities: An Interpretive Essay. *J. Economic Literature* 9:1–28.

Mishra, H.R. 1984. A Delicate Balance: Tigers, Rhinoceros, Tourists and Park Management vs. The Needs of the Local People in Royal Chitwan National Park, Nepal. In McNeely, J.A. and K.R. Miller (eds.). *National Parks, Conservation and Development: The Role of Protected Areas in Sustaining Society.* Smithsonian Institution Press, Washington, D.C.

Myers, N. 1983. *A Wealth of Wild Species.* Westview Press, Boulder, CO. 272 pp.

Myers, N. 1984. *The Primary Source.* W.W. Norton & Co., New York. 399 pp.

Myers, N. 1988. Tropical Forests: Much More than Stocks of Wood. *J. Tropical Ecology* 4:209–221.

Nations, J.D. and D.I. Komer, 1982. Indians, Immigrants, and Beef Exports: Deforestation in Central America. *Cultural Survival Quarterly* 6(2):8–12.

Newcombe, N. 1984. The Economic Justification for Rural Afforestation: The Case of Ethiopia. World Bank Energy Department Paper 16, Washington, D.C.

Newman, James R. and R. Kent Schereiber. 1984. Animals as Indicators of Ecosystem Responses to Air Emissions. *Environmental Management* 8(4):309–324.

Nicholls, Yvonne I. (Compiler). 1973. Source Book: Emergence of Proposals for Recompensing Developing Countries for Maintaining Environmental Quality. *IUCN Environmental Policy and Law Paper 5.* IUCN, Gland, Switzerland.

Norbu, Lhakpa Sherpa. 1987. Conservation and Management of Biological Resources in Protected Areas with Indigenous People. MS. 46 pp.

Norgaard, R.B. *et al.* 1984. The Economics of Cattle Ranching in Eastern Amazonia. University of California, Agricultural Experiment Station Working Paper 332:1–17.

Norgaard, R.B. 1984. *Environmental Economics: An Evolutionary Critique and a Plea for Pluralism.* Division of Agricultural Sciences, University of California, Berkeley, CA. Working Paper 299:1–24.

Norgaard, R.B. 1987. The Economics of Biological Diversity: Apologetics or Theory? In: Southgate, D.D. and J.F. Disinger (eds.). 1987.

Sustainable Resource Development in the Third World. Westview Press, Boulder, CO.

Norgaard, R.B. 1988. The Rise of the Global Exchange Economy and the Loss of Biological Diversity. pp. 206–211 in Wilson, E.O. (ed.). *Biodiversity.* National Academy Press, Washington, D.C.

Norton, Bryan. 1983. On the Inherent Danger of Undervaluing Species. Manuscript, Center for Philosophy and Public Policy, College Park, MD.

Norton, Bryan. 1985. Agricultural Development and Environmental Policy: Conceptual Issues. *Agriculture and Human Values* (Spring):63–70.

Norton, Bryan. 1986. *The Preservation of Species: The Value of Biological Diversity.* Princeton University Press, Princeton, NJ.

Norton, Bryan. 1988. Commodity, Amenity, and Morality: The Limits of Quantification in Valuing Biodiversity. pp. 200–205 in Wilson, E.O. (ed.). *Biodiversity.* National Academy Press, Washington, D.C.

OECD. 1982. *Economic and Ecological Interdependence.* Organization for Economic Cooperation and Development, Paris.

Oldfield, Margery. 1984. *The Value of Conserving Genetic Resources.* US Department of Interior, National Park Service. Washington, D.C. 360 pp.

Oldfield, Sara. 1988. Buffer Zone Management in Tropical Moist Forests. *IUCN Tropical Forest Paper* 5:1–49.

Organization of American States. 1987. *Minimum Conflict: Guidelines for Planning the Use of American Humid Tropic Environments.* OAS, Washington, D.C. 198 pp.

OTA (US Congress, Office of Technology Assessment). 1980. *Energy From Biological Processes.* Westview Press, Boulder, CO.

OTA (US Congress, Office of Technology Assessment). 1987. *Technologies to Maintain Biological Diversity.* U.S. Government Printing Office, Washington, D.C. 334 pp.

Page, T. 1977. *Conservation and Economic Efficiency.* Resources for the Future/Johns Hopkins University Press, Baltimore, MD.

Page, T and D. Maclean. 1983. Risk Conservatism and the Circumstances of Utility Theory. *American J. Agricultural Economics* 65(5):

Painter, Michael. 1988. Co-management With Whom? The Politics of Conservation and Development in Latin America. Paper presented at Smithsonian Institution Symposium on Culture and Conservation, April 8–9.

Panayotou, Theodore. 1987. Economics, Environment, and Development. *Harvard Institute for International Development* 259:1-29.

Panwar, H.S. 1980. Conservation-oriented development for Communities in Forested Regions of India. pp. 467-474. In *Tropical Ecology and Development*. Proc. Vth International Symposium on Tropical Ecology, Kuala Lumpur.

Partridge, E. (ed.) 1981. *Responsibilities to Future Generations: Environmental Ethics*. Prometheus Books, Buffalo, NY. 319 pp.

Pearce, D.W. 1975. *The Economics of Natural Resource Depletion*. Macmillan, London.

Pearce, D.W. 1976. *Environmental Economics*. Longmans, London.

Pearce, D.W. 1987a. The Sustainable Use of Natural Resources in Developing Countries. In R.K. Turner (ed). *Sustainable Environmental Management; Principles and Practice*. Frances Pinter, London.

Pearce, D.W. 1987b. Economic Values and the Natural Environment. *University College London Discussion Papers in Economics* 87(8): 1-20.

Pearce, D.W. 1987c. Marginal Opportunity Cost as a Planning Concept in Natural Resource Management. *University College London Discussion Papers in Economics* 87(6): 1-21.

Pearce, D.W. 1988. The Economics of Natural Resource Degradation in Developing Countries. pp. 102-117 in Turner, R.K. (ed.). *Sustainable Environmental Management: Principles and Practice*. Belhaven Press, London.

Pearce, D.W. and Nash, C.A. 1981. *Social Appraisal of Projects: A text in Cost-Benefit Analysis*. Macmillan, London. 225 pp.

Pearce, D. W. and A. Markandya. 1987. Marginal Opportunity Cost as a Planning Concept in Natural Resource Management. *Annals of Regional Science* 21(3):18-32.

Pearse, P.H. 1968. A New Approach to the Evaluation of Non-priced Recreational Resources. *Land Economics* 44:87-99.

Pearsall, S. 1984. In Absentia Benefits of Natural Preserves: A Review. *Environmental Conservation* 11(1):3-10.

Peskin, H.M. 1981. National Income Accounts and the Environment. *Natural Resources J.* 21:511-537.

Perrings, C.A. 1987. *Economy and Environment: A Theoretical Essay on the Interdependence of Economic and Environmental Systems*. Cambridge University Press, New York. 192 pp.

Perrings, C.A. 1988. An Optimal Path to Extinction? Poverty and Resource Degradation in the Open Agrarian Economy. *J. Development Economics* (in press).

Perrings, Charles, et al. 1988. *Economics and the Environment: A Contribution to the National Conservation Strategy for Botswana.* IUCN, Gland. 171 pp.

Peterson, George L. and Alan Randall. 1984. *Valuation of Wildlife Resource Benefits.* Westview Press, Boulder, CO. 258 pp.

Plourde, C. 1975. Conservation of Extinguishable Species. *Natural Resources J.* 15:791–798.

Pimlott, Douglas H. 1969. The Value of Diversity. *Trans. N. Am. Wildl. and Nat. Res. Conf.* 34:265–273.

Pindyck, Robert S. 1978. The Optimal Exploration and Production of Nonrenewable Resources. *J. Political Economy* 86(5): 841–861.

Pister, E.P. 1979. Endangered Species: Costs and Benefits. *Environmental Ethics.* 1:341–352.

Poore, Duncan, and J. Sayer. 1987. *The Management of Tropical Moist Forest Lands: Ecological Guidelines.* IUCN, Gland. 63 pp.

Porter, R.C. 1982. The New Approach to Wilderness Preservation Through Benefit-Cost Analysis, *J. Environmental Economics and Management* 9:59–80.

Prance, G.T. *et al.* 1987. Quantitative Ethnobotany and the Case for Conservation in Amazonia. *Conservation Biology* 1(4):296–310.

Praween Payapvipapong, Tavatchai Traitongyoo, R.J. Dobias. 1988. Using Economic Incentives to Integrate Park Conservation and Rural Development in Thailand. Paper presented at Workshop on Economics, IUCN General Assembly, 4–5 February 1988, Costa Rica.

Prescott-Allen, C. and R. Prescott-Allen. 1986. *The First Resource: Wild Species in the North American Economy.* Yale University Press, New Haven, CT. 529 pp.

Prescott-Allen, R. 1986. National Conservation Strategies and Biological Diversity. Report to IUCN, Gland, Switzerland.

Prescott-Allen, R. and C. Prescott-Allen. 1982. *What's Wildlife Worth? Economic Contributions of Wild Plants and Animals to Developing Countries.* International Institute for Environment and Development (Earthscan), London. 92 pp.

Principe, Peter P. 1988. Valuing Diversity of Medicinal Plants. Paper presented at IUCN/WHO/WWF International Consultation on the Conservation of Medicinal Plants, Chiangmai, Thailand.

Principe, Peter P. 1988. *The Economic Value of Biological Diversity Among Medicinal Plants.* OECD, Paris.

Ragozin, D.L. and G. Brown, Jr. 1985. Harvest Policies and Non-market Valuation in a Predator-Prey System. *J. Environmental Economics and Management* 12:155–168.

Randall, Alan and Hohn R. Stoll. 1983. Existence Value in a Total Valuation Framework. In Row, Robert D. and Lauraine G. Chestnut, (eds). *Managing Air Quality and Scenic Resources at National Parks and Wilderness Areas.* Westview Press, Boulder, CO. 314 pp.

Randall, Alan. 1979. *Resource Economics: An Economic Approach to Natural Resource and Environmental Policy.* Grid Publishing, Columbus, Ohio. 321 pp.

Randall, A. 1986. Human Preferences, Economics, and the Preservation of Species. pp. 79–109 in B.G. Norton, (ed.). *The Preservation of Species.* Princeton University Press, Princeton, N.J.

Randall, Alan. 1988. What Mainstream Economists Have to Say About the Value of Biodiversity. pp. 217–223 in Wilson, E.O. (ed). *Biodiversity.* National Academy Press, Washington D.C.

Ray, Anandarup. 1984. *Cost-Benefit Analysis: Issues and Methodologies.* Johns Hopkins University Press, Baltimore, MD.

Rendel, J. 1975. The Utilization and Conservation of the World's Animal Genetic Resources. *Agriculture and Environment* 2(2):101–119.

Repetto, Robert. 1986. Economic Policy Reform for Natural Resource Conservation (draft). World Resources Institute, Washington, D.C.

Repetto, Robert. 1987a. Creating Incentives for Sustainable Forest Development. *Ambio* 16(2–3):94–99.

Repetto, Robert. 1987b. Economic Incentives for Sustainable Production. *Annals of Regional Science* 21(3):44–59.

Repetto, Robert. 1988. *The Forest for the Trees? Government Policies and the Misuse of Forest Resources.* World Resources Institute, Washington, D.C. 105 pp.

Repetto, Robert, and William B. Magrath. 1988. *Natural Resources Accounting.* World Resources Institute, Washington, D.C.

Ricklefs, R.E., Z. Naveh, and R.E. Turner. 1984. Conservation of Ecological Processes. *IUCN Commission on Ecology Papers* 8:1–16.

Risbrudt, Christopher D., H.F. Kaiser, and Paul V. Ellefson. 1983. Cost-Effectiveness of the 1979 Forestry Incentives Programme. *J. Forestry.* 83(5):298–300.

Roberts, J.O.M. and B.D.G. Johnson. 1985. "Adventure" Tourism and Sustainable Development: Experience of the Tiger Mountain

Group's Operations in Nepal. pp. 81–84 in McNeely, J.A. and J.W. Thorsell (eds.). 1985. *People and Protected Areas in the Hindukush-Himalaya.* ICIMOD, Kathmandu. 250 pp.

Ronsivalli, L.J. 1978. Sharks and Their Utilization. *Marine Fisheries Review* 40(2):1–13.

Rolston, Holmes, III. 1985. Valuing Wildlands. *Environmental Ethics* 7:23–48.

Ruddle, K. 1973. The Human Use of Insects: Examples from the Yukpa. *Biotropica* 5(2):94–101.

Ruddle, K. 1986. No Common Property Problem: Village Fisheries in Japanese Coastal Waters. Paper presented at Workshop on Ecological Management of Common Property Resources, IV International Congress of Ecology, Syracuse, New York.

Ruggieri, G.D. and N.D. Rosenberg. 1978. *The Healing Sea.* Dodd Mead and Co., New York.

Runge, C. Ford. 1986. Common Property and Collective Action in Economic Development. *World Development* 14(5):623–35.

Sagoff, M. 1983. *Ethics and Economics in Environmental Policy and Planning.* Office of Environmental and Scientific Affairs, World Bank, Washington, D.C.

Sagoff, M. 1988. Some Problems with Environmental Economics. *Environmental Ethics* 10(1):55–74.

Sale, J.B. 1981. *The Importance and Values of Wild Plants and Animals in Africa.* IUCN, Gland, Switzerland. 44 pp.

Samuelson, P.A. 1976. Economics of Forestry in an Evolving Society. *Economic Inquiry* 14:466–492.

Samples, Karl, John Dixon, and Marcia Gowen. 1986. Information Disclosure and Endangered Species Evaluation. *Land Economics* 62(2):306–312.

Savina, Gail C. and Alan T. White. 1986. A Tale of Two Islands: Some Lessons for Marine Resource Management. *Environmental Conservation* 13(2): 107–113.

Scheuer, P.J. 1973. *Industry of Marine Natural Products.* Academic Press, New York.

Schmalensee, R. 1972. Option Demand and Consumer's Surplus: Valuing Price Changes Under Uncertainty. *American Economic Review* 62(5):813–824.

Schneider-Sawiris, Shadia. 1973. The Concept of Compensation in the Field of Trade and Environment. *IUCN Environmental Policy and Law Paper* 4:1–37.

Schonewald-Cox, Christine, *et al.* 1983. *Genetics and Conservation: A Reference for Managing Wild Animal and Plant Populations.* Benjamin/Cummings Publishing, Menlo Park, CA. 722 pp.

Schramm, G. 1985. *Practical Approaches to Estimating Resource Depletion.* Energy Department, World Bank, Washington, DC.

Schultes, R.E. and T. Swain. 1976. The Plant Kingdom: A Virgin Field for New Biodynamic Constituents. pp. 133–171 in N.J. Finer (ed.). *The Recent Chemistry of Natural Products, Including Tobacco: Proceedings of the Second Philip Morris Science Symposium.* Philip Morris, Inc., New York.

Schultze, William D., et al. 1983. The Economics of Preserving Visibility in the National Parklands of the Southwest. *Natural Resources J.* 23: 149–173.

Schulze, W., R. D'Arge, and D. Brookshire. 1981. Valuing Environmental Commodities: Some Recent Experiments. *Land Econ.* 151–171.

Schumann, Debra A. and William L. Partridge. 1986. *The Human Ecology of Tropical Land Settlement in Latin America.* Westview Press, Boulder CO.

Seidensticker, J. 1984. *Managing Elephant Depredation in Agriculture and Forestry Development Projects.* World Bank Technical Paper, Washington, D.C. 33 pp.

Sen, A.K. 1967. Isolation, Assurance, and the Social Rate of Discount. *Q.J. Econ.* 81:112–124.

Sevilla, Roque Larrea. 1988. Debt Swap for Conservation: The Ecuadorean Case. Paper presented at Workshop on Economics, IUCN General Assembly, 4–5 February 1988, Costa Rica.

Shane, D.R. 1986. *Hoofprints on the Forest: Cattle Ranching and the Destruction of Latin America's Tropical Forests.* Philadelphia Institute for the Study of Human Issues, 159 pp.

Shaw, W.W. and E.H. Zube (eds.). 1980. *Wildlife Values.* University of Arizona, School of Renewable Natural Resources, Tucson, AZ.

Sinden, J. 1967. The Evaluation of Extra-Market Benefits: A Critical Review. *World Agricultural Economics and Rural Sociology Abstracts* 9(4):1–16.

Sinden, J. 1981. Estimating the Value of Wildlife for Preservation: A Comparison of Approaches. *J. Environmental Management* 12:11–125.

Sinden, J. and A. Worrell. 1979. *Unpriced Values: Decisions Without Market Prices.* J. Wiley and Sons, New York.

Smith, R.J. 1971. The Evaluation of Recreation Benefits. *Urban Studies* 8:89–102.

Smith, V.K. (ed.). 1979. *Scarcity and Growth Reconsidered.* Johns Hopkins University Press, Baltimore, MD.

Smith, V.K. 1985. Intrinsic Values in Benefit Cost Analysis. USDA Forest Service, Rocky Mountain Forest and Range Experiment Station.

Smith, V.K. 1988. Resource Evaluation at the Crossroads. *Resources* 90:2–6.

Smith, V.K. and J.V. Krutilla. 1979. Endangered Species, Irreversibilities and Uncertainty: A Comment. *American J. Agricultural Economics* 61:371–375.

Smith, V.K. and J.V. Krutilla, J.V. (eds.). 1982. *Explorations in Natural Resources Economics.* Resources for the Future, Washington, D.C. 352 pp.

Soderbaum, P. 1980. Towards a Reconciliation of Economics and Ecology. *Eur. Rev. Agric. Econ.* 7:55–77.

Soderbaum, P. 1985. Economics, Evaluation and Environment. In: Hall, D.O., N. Myers and N.S. Margaris (eds.). *Economics of Ecosystem Management.* W. Junk Publishers, Dordrecht, The Netherlands.

Solow, R.M. 1974. The Economics of Resources and the Resources of Economics. *American Economic Review* 64:1–14.

Sorensen, John H., J. Soderstrom, and S.A. Carnes. 1984. Sweet for the Sour: Incentives in Environmental Mediation. *Environmental Management* 8(4):287–294.

Soulé, M.E. and B.A. Wilcox. 1980. *Conservation Biology.* Sinauer Associates, Sunderland, Massachusetts. 395 pp.

Spence, A.M. 1984. Blue Whales and Applied Control Theory. pp 43–71 in Y. Ahmad, P. Dasgupta and K-G Mler (eds.). *Environmental Decision Making.* Hodder and Stoughton, London.

Starkie, D.N.M. and D.M. Johnson. 1975. *The Economic Value of Peace and Quiet.* Heath, Lexington, MA.

Stevens, Joe B. 1969. Measurement of Economic Values in Sport Fishing: an Economist's Views on Validity, Usefulness, and Propriety. *Trans. Am. Fish. Soc.* 98(2):352–357.

Stoll, J.R. and L.A. Johnson. 1984. Concepts of Value, Nonmarket Values, and the Case of the Whooping Crane. Trans. *North Am. Wildl. Nat. Resour. Conf.* 49:382–393.

Stone, C. 1972. Should Trees Have Standing? Towards Legal Rights for Natural Objects. *Southern California Law Review* 45:450–501.

Sun, M. 1988. Costa Rica's Campaign for Conservation. *Science* 239:1366-1369.

Schwartzman, Stephan. 1987. Extractive Production in the Amazon Rubber Tappers' Movement. Paper presented to "Forests, Habitats, and Resources: A Conference in World Environmental History," 30 April, Duke University, Durham, NC.

Tarrant, James, *et al.* 1987. *Natural Resources and Environmental Management in Indonesia: An Overview.* USAID, Jakarta. 58 pp.

Thompson, D.N. 1973. *The Economics of Environmental Protection.* Winthrop Publishers, Inc., Cambridge, MA.

Tisdell, C.A. 1972. Provision of Parks and the Preservation of Nature: Some Economic Factors. *Australian Economic Papers* 11:154-162.

Tisdell, C.A. 1982. *Wild pigs: Economic Resource or Environmental Pest?* Pergamon Press, Sydney, Australia.

Tisdell, C.A. 1985a. Economics, Ecology, Sustainable Agricultural Systems and Development. *Development Southern Africa* 2(4):

Tisdell, C.A. 1985b. Sustainable Development: Conflicting Approaches of Ecologists and Economists, and Implications for LDCs. Occasional Paper No. 112. Department of Economics, University of Newcastle, N.S.W. 2308, Australia.

Trainer, D.O. 1973. Wildlife as Monitors of Disease. *American Journal of Public Health* 63(3):201-203.

Ulph, A.M. and I.K. Reynolds. 1981. *An Economic Evaluation of National Parks.* Centre for Resource and Environmental Studies, Australian National University, Canberra. 221 pp.

USAID. 1987. *AID Manual for Project Economic Analysis.* USAID Bureau for Program and Policy Coordination, Washington, D.C. 207 pp.

U.S. Water Resources Council. 1983. *Economic and Environmental Principles and Guidelines for Water and Related Land Resources Implementation Studies.* U.S. Government Printing Office, Washington, D.C.

USFWS. 1985. *Human Use and Economic Evaluation (HUEE).* ESM 104, Dept. of the Interior, Washington, D.C.

van Lavieren, L.P. 1983. *Wildlife Management in the Tropics with Special Emphasis on South-East Asia: A Guidebook for the Warden.* Handbook prepared for Ciawi School of Environmental Conservation Management. Bogor, Indonesia. 3 Vols.

Velozo, Ronnie de Camino. 1987. Incentives for Community Involvement in Conservation Programmes. *FAO Conservation Guide* 12:1-159.

Walsh, Richard G., J.B. Loomis, and R.A. Gillman. 1984. Valuing Option, Existence, and Bequest Demands for Wilderness. *Land Economics* 60(1):14–19.

Walsh, Richard G. and Lynde O. Gilliam. 1982. Benefits of Wilderness Expansion with Excess Demand for Indian Peaks. *Western J. Agricultural Economics* :1–12.

Warford, J. 1987a. *Environment, Growth and Development.* Economic Development Committee, World Bank, Washington D.C.

Warford, J. 1987b. Nature Resource Management and Economic Development. pp. 71–85 in Jacobs, P. and D. Munro (eds). *Conservation With Equity: Strategies for Sustainable Development.* IUCN, Gland, Switzerland. 466 pp.

Warford, J. 1987c. Natural Resources and Economic Policy in Developing Countries. *Annals of Regional Science* 21(3):3–17.

Weisbrod, B.A. 1964. Collective-Consumption Services of Individual Consumption Goods. *Quarterly J. Economics* 78(2):471–477.

Wiess, E.B. 1984. The Planetary Trust: Conservation and Intergenerational Equity. *Ecology Law Quarterly* 11(4):495–582.

Wennergren, E. Boyd. 1967. Surrogate Pricing of Outdoor Recreation. *Land Econ.* 43:112–115.

Western, D. 1984. Amboseli National Park: Human Values and the Conservation of a Savanna Ecosystem. In McNeely, J.A. and K.R. Miller (eds). *National Parks, Conservation, and Development: The Role of Protected Areas in Sustaining Society.* Smithsonian Institution Press, Washington, D.C.

Western, D. and W. Henry. 1979. Economics and Conservation in Third World National Parks. *Bioscience* 29(7):414–418.

Westman, Walter E. 1977. How much are nature's services worth? *Science* 197: 960–964.

White, Alan and Dale Law. 1986. Evaluation of the Marine Conservation and Development Program of Silliman University, Philippines. *MCDP Newsletter* 6:1–15.

Wilson, E. (ed.). 1988. *Biodiversity.* National Academy Press, Washington, D.C. 521 pp.

Wolverton, B.C. and R.C. McDonald. 1981. Natural Processes for Treatment of Organic Chemical Waste. *The Environmental Professional* 3:99–104.

World Bank. 1986. *Wildlands: Their Protection and Management in Economic Development.* World Bank, Washington, D.C. 220 pp.

World Commission on Environment and Development. 1987. *Our Common Future.* Oxford University Press, Oxford, UK.

Worster, Donald. 1985. *Nature's Economy.* Cambridge University Press, Cambridge, UK.

WRI/IIED. 1986. *World Resources 1986: An Assessment of the Resource Base that Supports the Global Economy.* Basic Books, Inc., New York. 353 pp.

LIST OF ACRONYMS

BOSTID Board on Science and Technology for International Development
CASDC Committee on Agricultural Sustainability for Developing Countries (USA)
CATIE Centro Agronómico Tropical de Investigación y Ensenanza (Tropical Agricultural Research and Training Center, Costa Rica)
CEMP Centre for Environmental Management and Planning
CIDA Canadian International Development Agency
CITES Convention on International Trade in Endangered Species of Wild Fauna and Flora
CPI Acre Pro-Indian Commission (Brazil)
EEC European Economic Community
EIA Environmental Impact Assessment
EPS Environmental Protection Society (Thailand)
ESA Environmentally Sensitive Area (United Kingdom)
FAO Food and Agriculture Organization of the United Nations
FCA Fisheries Cooperative Association (Japan)
GDP Gross Domestic Product
GNP Gross National Product ha Hectare
IBPGR International Board for Plant Genetic Resources
IIED International Institute for Environment and Development
IPAL Integrated Project on Arid Lands (Kenya)
ITTO International Tropical Timber Organization
IUCN International Union for Conservation of Nature and Natural Resources
MOC Marginal Opportunity Cost
MMC Marine Management Committee (Philippines)
NCS National Conservation Strategy

NGO	Non-Governmental Organization
NORAD	Norwegian Agency for International Development
NPWD	National Parks and Wildlife Department (Zambia)
OTA	Office of Technology Assessment of the US Congress
PDA	Population and Community Development Association (Thailand)
PCMB	Porcupine Caribou Management Board (Canada)
SAED	Special Areas for Eco-Development (India)
SIDA	Swedish International Development Authority
UK	United Kingdom of Great Britain and Ireland
UN	United Nations
UNEP	United Nations Environment Programme
UNESCO	United Nations Education, Scientific, and Cultural Organization
USA	United States of America
USAID	United States Agency for International Development
USFWS	United States Fish and Wildlife Service
USNPS	United States National Park Service
WCED	World Commission on Environment and Development
WCMC	World Conservation Monitoring Centre
WCS	World Conservation Strategy
WRI	World Resources Institute
WWF	Worldwide Fund for Nature (previously World Wildlife Fund, and still World Wildlife Fund in some countries)

INDEX